ANNALS OF THE NEW YORK ACADEMY OF SCIENCES

Volume 536

EDITORIAL STAFF
Executive Editor
BILL BOLAND
Managing Editor
JUSTINE CULLINAN
Associate Editors
TRUMBULL ROGERS
JOYCE HITCHCOCK

The New York Academy of Sciences
2 East 63rd Street
New York, New York 10021

THE NEW YORK ACADEMY OF SCIENCES
(Founded in 1817)

BOARD OF GOVERNORS, 1988

WILLIAM T. GOLDEN, *President*
LEWIS THOMAS, *President-Elect*

Honorary Life Governors

SERGE A. KORFF	H. CHRISTINE REILLY	IRVING J. SELIKOFF

Vice-Presidents

CYRIL M. HARRIS	PIERRE C. HOHENBERG	DENNIS D. KELLY
PETER D. LAX		JAMES G. WETMUR

ALAN PATRICOF, *Secretary-Treasurer*

Elected Governors-at-Large

THORNTON F. BRADSHAW	DAVID A HAMBURG	GERALD D. LAUBACH
NEAL E. MILLER	JOSEPH F. TRAUB	ERIC J. SIMON

Past Presidents (Governors)

WILLIAM S. CAIN	FLEUR L. STRAND

HEINZ R. PAGELS, *Executive Director*

INTEGRABILITY IN DYNAMICAL SYSTEMS

ANNALS OF THE NEW YORK ACADEMY OF SCIENCES
Volume 536

INTEGRABILITY IN DYNAMICAL SYSTEMS

Edited by J. Robert Buchler, James R. Ipser, and Carol A. Williams

The New York Academy of Sciences
New York, New York
1988

Copyright © 1988 by the New York Academy of Sciences. All rights reserved. Under the provisions of the United States Copyright Act of 1976, individual readers of the Annals are permitted to make fair use of the material in them for teaching and research. Permission is granted to quote from the Annals provided that the customary acknowledgment is made of the source. Material in the Annals may be republished only by permission of the Academy. Address inquiries to the Executive Editor at the New York Academy of Sciences.

Copying fees: For each copy of an article made beyond the free copying permitted under Section 107 or 108 of the 1979 Copyright Act, a fee should be paid through the Copyright Clearance Center Inc., 21 Congress St., Salem, MA 01970. For articles of more than 3 pages the copying fee is $1.75.

Library of Congress Cataloging-in-Publication Data

Integrability in dynamical systems/edited by J. Robert Buchler, James R. Ipser, and Carol A. Williams.
   p. cm.—(Annals of the New York Academy of Sciences, ISSN 0077-8923; v. 536)
   Bibliography: p.
   Includes index.
   ISBN 0-89766-468-X. ISBN 0-89766-469-8 (pbk.)
   1. Astrophysics—Mathematics—Congresses. 2. Dynamics—Congresses. 3. Nonlinear theories—Congresses. I. Buchler, J. R. (J. Robert) II. Ipser, James R. III. Williams, Carol A. IV. Series.
Q11.N5 vol. 536
[QB460]
500 s—dc19
[523.01'0151]   88-22409   CIP

SP
Printed in the United States of America
ISBN 0-89766-468-X (cloth)
ISBN 0-89766-469-8 (paper)
ISSN 0077-8923

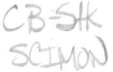

ANNALS OF THE NEW YORK ACADEMY OF SCIENCES

Volume 536

August 12, 1988

INTEGRABILITY IN DYNAMICAL SYSTEMS[a]

Editors
J. ROBERT BUCHLER, JAMES R. IPSER, and CAROL A. WILLIAMS

CONTENTS

Preface. *By* J. ROBERT BUCHLER, JAMES R. IPSER, and CAROL A. WILLIAMS	vii
Resonant Integrable Galactic Models. *By* G. CONTOPOULOS	1
Integrable Models for Galaxies. *By* TIM DE ZEEUW	15
Integrable Galactic Models. *By* C. HUNTER	25
New Integrable Systems. *By* JARMO HIETARINTA	33
Painlevé Expansions for Integrable and Nonintegrable Ordinary Differential Equations. *By* M. TABOR	43
Solving the Vlasov Equation in General Relativity by Particle Simulation. *By* STUART L. SHAPIRO and SAUL A. TEUKOLSKY	53
Repulsive and Attractive Double-Bubble Space–Times. *By* JAMES R. IPSER	77
Integrability of Magnetic Confinement Systems. *By* R. V. E. LOVELACE	81
Eternity, Chaos, Lie Algebras, Integrability, and Accelerator Design. *By* ALEX J. DRAGT	83
Hannay's Angle and Berry's Phase in the Classical Adiabatic Motion of Charged Particles. *By* ROBERT G. LITTLEJOHN	87
Integrability of Nonlinear Wave Equations. *By* H. H. CHEN and J. E. LIN	91
Normalization in the Face of Integrability. *By* ANDRÉ DEPRIT and BRUCE R. MILLER	101
Simplifications toward Integrability of Perturbed Keplerian Systems. *By* SEBASTIAN FERRER and CAROL A. WILLIAMS	127
Index of Contributors	141

[a]The papers in this volume were presented at the Third Florida Workshop in Nonlinear Astronomy, entitled Integrability in Dynamical Systems, which was held on October 1–2, 1987 in Gainesville, Florida.

The New York Academy of Sciences believes it has a responsibility to provide an open forum for discussion of scientific questions. The positions taken by the participants in the reported conferences are their own and not necessarily those of the Academy. The Academy has no intent to influence legislation by providing such forums.

Preface

J. ROBERT BUCHLER,[a] JAMES R. IPSER,[a]
AND CAROL A. WILLIAMS[b]

[a]*Department of Physics*
University of Florida
Gainesville, Florida 32611

[b]*Astronomy Program and Mathematics Department*
University of South Florida
Tampa, Florida 33620

This workshop on Integrability in Dynamical Systems, the third in a series of Florida Workshops in Nonlinear Astronomy, was held at the University of Florida on October 1 and 2, 1987. The common motivation of these workshops is to bring together an interdisciplinary group of scientists for a period of concentration on specialized research topics in nonlinear science of interest to Astronomy. Like the previous two workshops, namely the 1986 workshop on Orbits in Galaxies and the 1987 workshop on Chaotic Phenomena in Astrophysics, the latter published in the *Annals of the New York Academy of Sciences*, the 1988 workshop was organized under the leadership of Professor George Contopoulos, himself a pioneering figure in the study of integrable as well as chaotic systems. The following remarks are based largely on his introductory comments at the workshop and on Jim Ipser's concluding remarks.

The previous workshop dealt with chaotic systems, both Hamiltonian and dissipative. It is well known that in Hamiltonian systems, from a strict mathematical point of view, integrability is the exception rather than the rule. In practice it so happens, however, that a great deal of insight can be gained from the study of integrable systems, and this provides motivation for a workshop on a topic seemingly of measure-zero interest. The collection of articles in this volume certainly bears witness to the astrophysical interest of integrable systems.

First, integrability plays an important role in galactic dynamics. Although realistic galactic models have both ordered and stochastic (chaotic) orbits, the famous KAM (Kolmogorov–Arnold–Moser) theorem tells us that regular multiperiodic orbits can still be dominant in systems that are close to being integrable. As an example of a study that is specifically based on integrability we mention the well-known density wave theory of spiral structure. Lynden-Bell deserves much of the credit for introducing integrable models. However, recent years have seen a flurry of research in this area and the three articles in this volume by George Contopoulos, Tim DeZeeuw, and Chris Hunter report on recent progress.

A second astrophysical prototype of ordered motions is provided by our Solar System. Celestial Mechanics has dealt with approximately integrable systems for more than two centuries. No doubt some orbits in the Solar System are known to be chaotic. Nevertheless the theory and computation of ordered motions have reached unprecedented accuracy. André Deprit has advanced the field of lunar and planetary theory through his exploitation of algebraic computer methods, and here he describes the usefulness of such methods in analyses of various dynamical systems. In order to obtain the kind of accuracy that is now within reach of measurement the best perturbation techniques require so many terms as to be almost intractable. The paper by Sebastian

Ferrer and Carol Williams discusses ways of "simplifying" the problem beforehand with the use of canonical transformations.

A similar application of integrability appears in accelerators, although here the timescale problem is exacerbated by many orders of magnitude compared to galactic or even celestial mechanics problems. In order to avoid a fiasco in the next generation of accelerators, such as the superconducting supercollider, a large amount of theoretical work is being performed to avoid stochasticity. Alex Dragt, who is one of the leaders in this field, reports on the current status of accelerator design problems and the use of Lie algebra techniques to resolve some of these problems.

In plasma physics the situation is quite complicated. The role of collective phenomena there is more pronounced and the large number of instabilities add additional complexities. However, ordered motions do play an important role. Lovelace's abstract summarizes his paper describing the usefulness of various methods for solving plasma problems, depending on whether the orbits are ergodic or are constrained by extra integrals of the motion. Robert Littlejohn reexamines the classical adiabatic motion of a charged particle in a strong magnetic field and shows how a novel gauge transformation is associated with this phase.

From a more mathematical point of view, a happy surprise was the discovery of many new exactly integrable Hamiltonian systems. This work started with the proof of the integrability of the Toda lattice by Michel Hénon and Hermann Flashka in 1974 and was followed by a large number of papers by many people. Flashka reported about his recent work in this field. We are grateful to Jarmo Hietarinta for traveling from Finland especially for this workshop in order to review the status of invariants and of integrability in Hamiltonian systems with two degrees of freedom. Mike Tabor addresses the intriguing relation between integrability and properties of the solutions in the complex plane, viz., the so-called Painlevé property and its generalization.

An extension of the search for integrability in partial differential equations (PDE) opened new horizons of research. This field started with the work of Kruskal and Zabusky on the behavior of a numerical simulation of the Fermi–Pasta–Ulam system and their discovery of persistent nonlinear features, which were subsequently labeled solitons. The development of inverse scattering theory clearly showed the connection between solitons and the integrability of the PDEs. This subject could easily fill a large volume by itself. Here we get a feeling for its importance and scope from the paper by Dr. H.-H. Chen, who has himself contributed considerably in this area.

Another area where integrability was encountered is General Relativity. The proof of the integrability of the Kerr black hole by Carter came as a surprise. But this result has had immediate applications and extensions. The mathematical theory of black holes, as developed mainly by Chandrasekhar, has shown many more unexpected and surprising symmetries and regularities. Furthermore, new solutions of Einstein's equations have been found, especially in connection with Cosmology, and numerical techniques have been applied to solve problems that were otherwise very difficult. The paper by Stuart Shapiro and Saul Teukolsky discusses their novel attack on the problem of a collisionless gas of particles in general relativity. Jim Ipser, on the other hand, introduces us to some avant-garde integrable cosmological solutions involving topological defects known as domain walls.

Finally we would like to express our gratitude to Professor Steve Gottesman, without whose dedicated organizational skill this workshop would not have been the success it was.

Resonant Integrable Galactic Models

G. CONTOPOULOS

Department of Astronomy
University of Florida
Gainesville, Florida 32611
and
Department of Astronomy
University of Athens
Athens, Greece

INTRODUCTION

Integrable models have been used on various occasions in galactic dynamics. For example, Kuzmin,[1] Hori,[2] and van de Hulst[3] have used a Stäckel potential as a model of our Galaxy.

Recently, several useful applications of 2- and 3-dimensional integrable galactic models have been made by de Zeeuw and Lynden-Bell,[4] de Zeeuw[5] and de Zeeuw *et al.*[6]

However, most of these models (a) deal with nonrotating systems and (b) do not treat the main resonant phenomena in galaxies; still, such phenomena are important in realistic models of galaxies.[7]

In order to understand the resonance phenomena in rotating galaxies it is useful to introduce particular resonant integrable models (Contopoulos[8,9]). These models have the advantage that they put the emphasis on one particular resonance at a time and study its effects. On the other hand, their disadvantage is that these models are only local and not global. It was found that the theoretical results based on simple integrable models agree closely with the empirical results, found by numerical integration, if the perturbation is small. However, in cases of strong perturbations the numerical agreement is less good.

THE DYNAMICS OF RESONANCES

We summarize here the basic method used in treating the resonances in flat rotating galactic models. These models are of the form

$$V = V_0(r) + \epsilon V_1(r, \theta), \quad (1)$$

where $V_0(r)$ is the axisymmetric background potential and $\epsilon V_1(r, \theta)$ is the spiral, or bar perturbation. The Hamiltonian is

$$H = V - \Omega_s J_0, \quad (2)$$

where J_0 is the angular momentum and Ω_s the angular velocity of the spiral or bar perturbation, assumed to be constant. The Hamiltonian is a constant of motion (Jacobi constant, or energy in the rotating frame). We assume further that the model is bisymmetric, that is, the angle θ appears only in even multiples in a Fourier analysis of $V_1(r, \theta)$.

If we use action-angle variables, we can write

$$\epsilon V_1 = \sum_m \sum_n [\epsilon_{m,n} \cos(m\theta_1 - n\theta_2) + \epsilon'_{m,n} \sin(m\theta_1 - n\theta_2)], \quad (3)$$

where θ_1, is the "empicyclic angle," θ_2 is the azimuth of the "epicyclic center," and $n = 0, \pm 2, \pm 4, \ldots$.

A resonance appears when the epicyclic frequency ω_1 and the rotational frequency in the rotating frame ω_2 have a rational ratio

$$\frac{\omega_1}{\omega_2} = \frac{n}{m}. \quad (4)$$

In the case of a small perturbation, ω_1 is close to the epicyclic frequency, κ, and ω_2 is close to $(\Omega - \Omega_s)$, where Ω is the angular velocity at a distance r_c. The actions corresponding to the angles θ_1, θ_2 are the "epicyclic action," I_1, and the azimuthal action $I_2 = J_0 - J_c$, where J_c is the angular momentum of an unperturbed circular orbit of radius r_c.

A galactic model is "resonant integrable" if it contains only one combination of angles $(m\theta_1 - n\theta_2)$ and its multiples. As an example we consider the Hamiltonian

$$H \equiv \omega_1 I_1 + \omega_2 I_2 + aI_1^2 + 2bI_1I_2 + cI_2^2 + \epsilon_{m,n}(I_1, I_2) \cos(m\theta_1 - n\theta_2) = 0 \quad (5)$$

that is, we include only the most important terms containing the actions (i.e., the linear and quadratic terms) and use only one lowest order resonant trigonometric term. The numerical value of the Hamiltonian in this form is zero.[8]

A canonical change of variables

$$\psi_1 = \theta_1 - \frac{n}{m}\theta_2, \qquad I_1 = J_1$$

$$\psi_2 = \theta_2, \qquad I_2 = J_2 - \frac{n}{m}J_1 \quad (6)$$

brings the Hamiltonian to the form

$$H \equiv \left(\omega_1 - \frac{n}{m}\omega_2\right)J_1 + \omega_2 J_2 + aJ_1^2 + 2bJ_1\left(J_2 - \frac{n}{m}J_1\right)$$

$$+ c\left(J_2 - \frac{n}{m}J_1\right)^2 + \epsilon_{m,n} \cos m\psi_1 = 0. \quad (7)$$

As the angle ψ_2 is ignorable, the action J_2 is an integral of motion. This form of the Hamiltonian has been used to find the main resonance phenomena at resonances.[9] In the case of an even resonance

$$\frac{\omega_1}{\omega_2} = \frac{2n'}{1}, \quad (8)$$

the main resonance phenomenon is the appearance of a gap in the characteristic of the "central" family of periodic orbits (the orbits that are reduced to circles when the perturbation goes to zero), while in the case of an odd resonance

$$\frac{\omega_1}{\omega_2} = \frac{2(2n' + 1)}{2} \tag{9}$$

($m = 2$, because n has to be even), it is the appearance of an instability strip along the characteristic of the main family of periodic orbits. By using Cartesian coordinates

$$x = (2J_1)^{1/2} \cos \psi_1, \qquad \dot{x} = -(2J_1)^{1/2} \sin \psi_1, \tag{10}$$

we express H as a function of x and $\dot{x}$

$$H \equiv H(x, \dot{x}) = 0. \tag{11}$$

Here H also contains the constant parameters ω_1, ω_2, a, b, c, ϵ, that depend on r_c, and the integral J_2.

The periodic orbits are found by taking

$$\frac{\partial H}{\partial x} = \frac{\partial H}{\partial \dot{x}} = 0 \tag{12}$$

for every value of r_c. The three equations 11–12 give the values of x, $\dot{x}$, and J_2 as functions of the unperturbed radius r_c (or of the corresponding energy, h, of the unperturbed circular orbit).

Here we will use models somewhat simpler than in our previous paper,[9] and we will apply them to two representative even and odd resonances, namely $n/m = 4/1$ and $n/m = 6/2 = 3/1$.

RESONANCE 4/1

We assume

$$\epsilon_{4,1} = \epsilon(2J_1)^{1/2} \tag{13}$$

and find

$$H = \omega_2 J_2 + cJ_2^2 + Q\left[\frac{q}{2}(J_{20} - J_2)(x^2 + \dot{x}^2) + \frac{1}{4}(x^2 + \dot{x}^2)^2 + \epsilon_1 x\right] = 0, \tag{14}$$

where

$$Q = a - 8b + 16c,$$

$$qQ = -2(b - 4c),$$

$$qQJ_{20} = \omega_1 - 4\omega_2,$$

$$\epsilon_1 Q = \epsilon. \tag{15}$$

In general we have $Q < 0$ and $q > 0$. The various quantities change only slightly with r_c, except J_{20}, which is zero at exact resonance $r_c = r_{res}$ (when $\omega_1 = 4\omega_2$) and changes almost linearly with $(r_c - r_{res})$. If $r_c > r_{res}$ it is $J_{20} < 0$.

The periodic orbits satisfy the relations

$$\frac{\partial H}{\partial \dot{x}} = Q\dot{x}\,[q(J_{20} - J_2) + x^2 + \dot{x}^2] = 0,$$

$$\frac{\partial H}{\partial x} = Q\{\epsilon_1 + x[q(J_{20} - J_2) + x^2 + \dot{x}^2]\} = 0. \quad (16)$$

Thus we have $\dot{x} = 0$ and

$$x^3 + q(J_{20} - J_2)x + \epsilon_1 = 0. \quad (17)$$

It is obvious that x has either one or three values, depending on the values of the constants q, J_{20}, ϵ_1, and J_2.

If $\epsilon_1 = 0$ we have either $x = 0$ ("central" family) or $x^2 = q(J_2 - J_{20})$ (bifurcating family). In the first case, $J_2 = 0$, while in the second case

$$\omega_2 J_2 + c J_2^2 = \frac{Qq^2}{4}(J_2 - J_{20})^2. \quad (18)$$

The solution of this equation in the lowest order is

$$J_2 = \frac{Qq^2 J_{20}^2}{4\omega_2}, \quad (19)$$

that is, J_2 is of second order in J_{20}. Then in the lowest approximation

$$x^2 = -qJ_{20}; \quad (20)$$

hence, the bifurcation occurs at exact resonance ($J_{20} = 0$) and the bifurcating family exists for $r_c > r_{res}$.

If $\epsilon_1 \neq 0$, we solve equation 17 for J_2 and insert it in equation 14 with $\dot{x} = 0$ to find

$$x = -\frac{c\epsilon_1}{q\omega_2} + \frac{c^2\epsilon_1}{q\omega_2^2} J_{20} \quad (21)$$

in the lowest order in ϵ_1 and J_{20}. The first term is almost constant, while the second term has a sign opposite to ϵ_1 for $r_c > r_{res}$. Thus for $\epsilon_1 < 0$ ($\epsilon > 0$; case 1) the deviation x increases with r_c as in FIGURE 1a, while for $\epsilon_1 > 0$ ($\epsilon < 0$; case 2) the deviation x decreases with r_c, as in FIGURE 1b.

The transition from case 1 to case 2 is shown in FIGURE 1c. As ϵ decreases, the characteristics x_1 and $4/1$ approach each other, and for ϵ slightly negative they separate in a different way than for ϵ positive.

The corresponding forms of the orbits in cases 1 and 2 are given in FIGURES 2a and 2b. The first case corresponds to models given by Contopoulos and Papayannopoulos,[10]

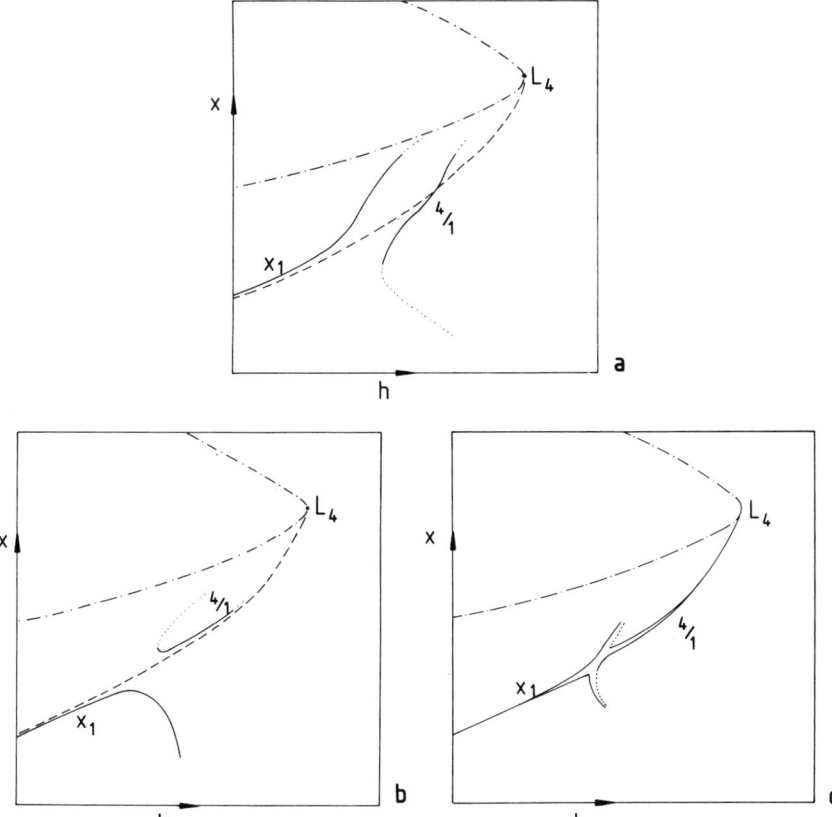

FIGURE 1. Characteristics of families of periodic orbits (distance x versus energy h in the rotating frame, schematically) near the 4/1 resonance. The *dashed line* represents the characteristic of the "central" family of circular orbits, when $\epsilon = 0$. It extends all the way from the center to the Lagrangian point L_4 at corotation. If $\epsilon \neq 0$, a gap separates the family x_1 from two 4/1 families, one stable and one unstable. The gap is of the form (**a**) if $\epsilon > 0$ and (**b**) if $\epsilon < 0$. The transition cases are shown in (**c**). *Solid lines* indicate stable orbits, and *dotted lines* unstable orbits. The *dash–dotted line* is the curve of zero velocity.

Teuben and Sanders,[11] Papayannopoulos and Petrou,[12] Contopoulos et al.,[13] and Papayannopoulos.[14] The second case corresponds to models given by Athanassoula et al.[15] and Pfenniger.[16]

Any bar model can be Fourier analyzed in the form

$$V = \sum_{n=0}^{\infty} A_{2n} \cos 2n\theta. \qquad (22)$$

By a convenient canonical transformation of the variables, we derive the 4/1 resonant terms of the form ${}_k\epsilon_{4,1} \cos k(\theta_1 - 4\theta_2)$. The most important terms (numerically) are in general $A_2 \cos 2\theta$ and $A_4 \cos 4\theta$. Both these components produce resonant terms of the

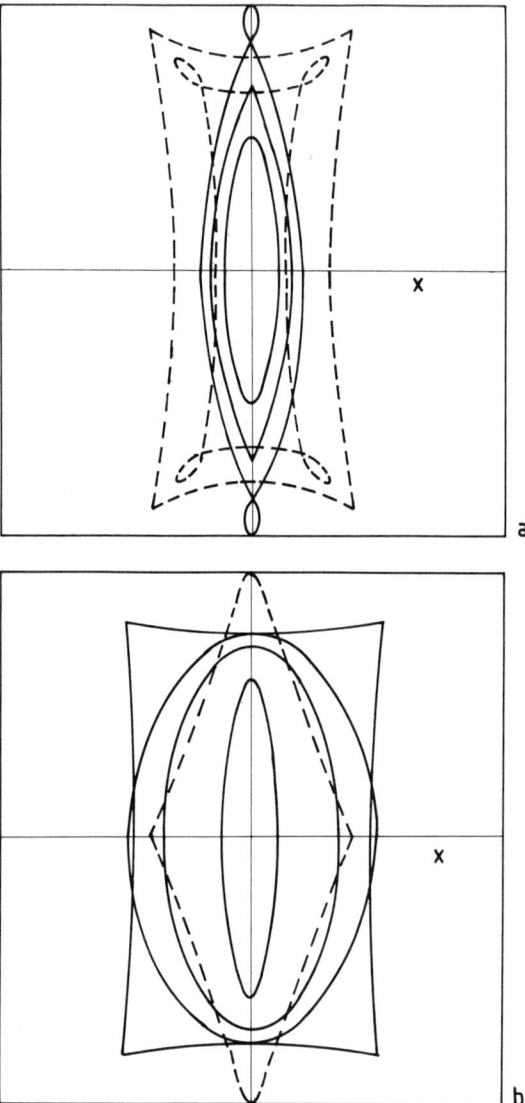

FIGURE 2. Orbits near the 4/1 resonance in cases (**a**) and (**b**) of FIGURE 1 (schematically). *Solid lines* represent the orbits of the family x_1, and *dashed lines* show the orbits of the family 4/1.

form $\epsilon_{4,1} \cos(\theta_1 - 4\theta_2)$ in the Hamiltonian.[17] However, the term $A_4 \cos 4\theta$ produces a *linear* term in A_4, while the term $A_2 \cos 2\theta$ produces a *quadratic* term [of $0(A_2^2)$]. The quadratic term is positive, while we take the linear term negative. Thus if we have a weak bar, the linear term dominates and gives $\epsilon < 0$, that is, a gap of type 2, while if we increase all $A_2, A_4, \ldots$ by a sufficiently large factor, the quadratic term becomes stronger and gives $\epsilon > 0$, that is, a gap of type 1.

We have recently studied[13] many models of the form

$$V = v_{max}^2 [\ln r + E_1 (0.242r)] + \frac{A}{r^2} e^{-0.4r} (0.79983r - 1) (\cos 2\theta + \alpha_4 \cos 4\theta) \quad (23)$$

(outside $r = 1.9523$), where E_1 is the exponential integral, $v_{max} = 280$, and the angular velocity of the bar is $\Omega_s = 52.4$, for various values of A and a_4. In the particular case $A = 97422$, $a_4 = 0$, such a model represents, in an approximate way, the galaxy NGC 3992. In the case $A = 10000$, the gap is of type 1 if $\alpha_4 \geq -0.04$, while it is of type 2 if $\alpha_4 \leq -0.05$.

When A is relatively small (e.g., $A = 10000$) the family 4/1 beyond the gap is only stable for large enough r_c (FIG. 3). Thus the orbits of this family play a role only in the outer parts of the bar. However, when A becomes larger ($A \gtrsim 75000$), the stable 4/1 family extends all the way to the center and plays an important role for the dynamics of the whole galaxy. In the case of the NGC 3992 model ($A = 97422$) the orbits of both families x_1 and 4/1 contribute in constructing models of the galaxy (FIG. 4). Thus the galaxy takes a boxy type of the same form as some observed galaxies (FIG. 5). Along the axis of the bar there are loops along the x_1 orbits that may produce particularly strong features near the end of the bar (FIGS. 4 and 5). A similar model with a boxy bar but stronger loops of the x_1 family was provided in plate 1 of Teuben and Sanders.[11]

RESONANCE 3/1

In this case we have $n/m = 3/1$, and we must take $m = 2$ and $n = 6$, because n is even. The perturbation is

$$\epsilon_{6,2} \cos (2\theta_1 - 6\theta_2), \quad (24)$$

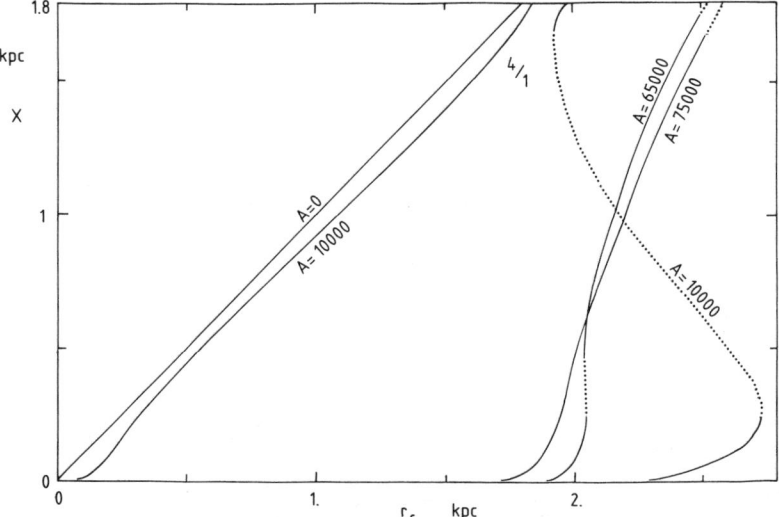

FIGURE 3. Characteristics of the families x_1 and 4/1 in weak and strong bars. In strong enough bars the family 4/1 is stable from the center up to almost the corotation distance.

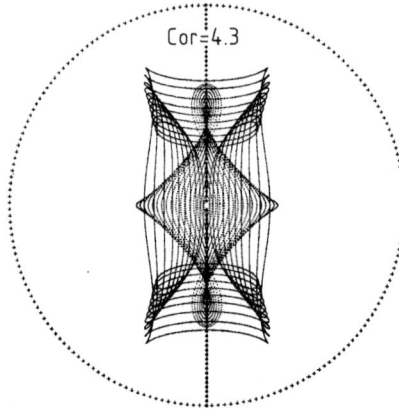

FIGURE 4. Stable periodic orbits supporting the bar in a strong bar model (equation 23).

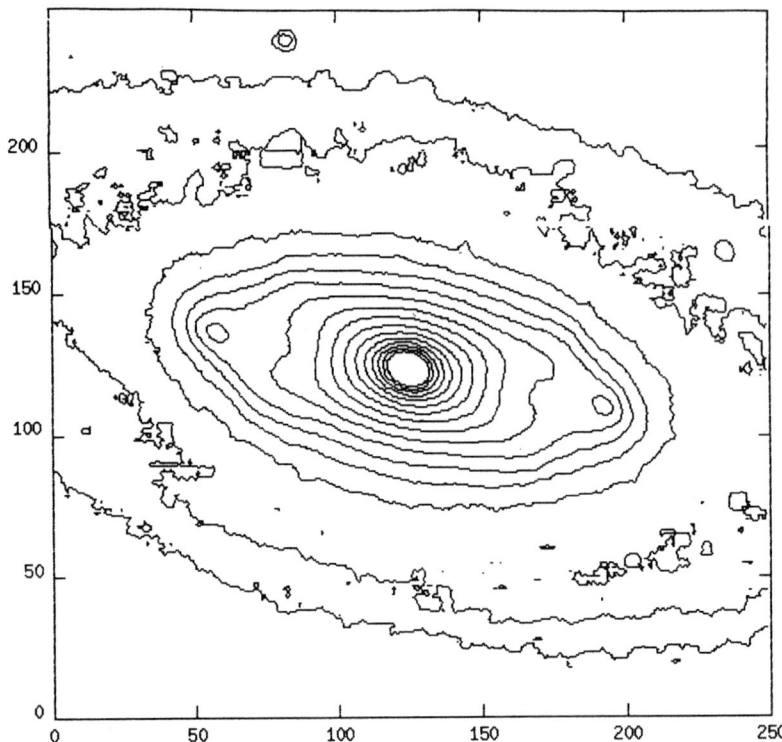

FIGURE 5. The isophotes of the bar region of the galaxy NGC 4321 have a characteristic boxy form. The outermost hexagonal isophote indicates that there is also an appreciable amount of x_1 orbits with loops beyond the 4/1 resonance. (Figure provided by R. Buta.)

and we take for $\epsilon_{6,2}$ the simple form

$$\epsilon_{6,2} = \epsilon(2J_1). \tag{25}$$

Then the Hamiltonian is written

$$H \equiv \omega_2 J_2 + cJ_2^2 + Q\left[\frac{q}{2}(J_{20} - J_2)(x^2 + \dot{x}^2)\right.$$
$$\left. + \frac{1}{4}(x^2 + \dot{x}^2)^2 + \epsilon_1(x^2 - \dot{x}^2)\right] = 0, \tag{26}$$

where

$$Q = a - 6b + 9c,$$
$$qQ = -2(b - 3c),$$
$$qQJ_{20} = \omega_1 - 3\omega_2,$$
$$\epsilon_1 Q = \epsilon. \tag{27}$$

Then equations 12 give for the periodic orbits

$$x[q(J_{20} - J_2) + x^2 + \dot{x}^2 + 2\epsilon_1] = 0,$$
$$\dot{x}[q(J_{20} - J_2) + x^2 + \dot{x}^2 - 2\epsilon_1] = 0. \tag{28}$$

There are three solutions of this system

(a) $\quad x = \dot{x} = 0,$

(b) $\quad \dot{x} = 0, \quad x^2 = q(J_2 - J_{20}) - 2\epsilon_1, \tag{29}$

(c) $\quad x = 0, \quad \dot{x}^2 = q(J_2 - J_{20}) + 2\epsilon_1. \tag{30}$

In case (a) we find from equations 10 and 26, $J_1 = J_2 = 0$. This solution represents exactly the "central" family of circular periodic orbits of the unperturbed problem. This special form of the orbits is due to the particular choice of $\epsilon_{6,2}$. In more general cases the solutions deviate from circles.

In cases (b) and (c) we insert the values of x and $\dot{x}$ in equation 26 and find in the lowest approximation

$$J_2 = \frac{Q}{4\omega_2}(qJ_{20} + 2\epsilon_1)^2. \tag{31}$$

Thus J_2 is of second order in J_{20} and ϵ_1. Keeping only terms of first order in J_{20} and ϵ_1 in equations 29 and 30, we find

$$x^2 = -qJ_{20} - 2\epsilon_1, \tag{32}$$
$$\dot{x}^2 = -qJ_{20} + 2\epsilon_1. \tag{33}$$

Thus we have two 3/1 bifurcating families, one symmetric with respect to the x axis ($\dot{x} = 0$) and one asymmetric ($\dot{x} \neq 0$). The characteristics are like parabolas, starting on the $x = \dot{x} = 0$ characteristic, close to $J_{20} = 0$ ($\omega_1 - 3\omega_2 = 0$). Both parabolas open in the direction of increasing r_c if $Q < 0$ (because $-qJ_{20} = -(\omega_1 - 3\omega_2)/Q$), or in the direction of decreasing r_c, if $Q > 0$. The bifurcations occur at $\omega_1 - 3\omega_2 = \mp 2\epsilon$. If $\epsilon > 0$, the symmetric family starts at $r_c < r_{res}$ and the asymmetric one at $r_c > r_{res}$, and vice versa if $\epsilon < 0$. All the combinations of relative positions and directions of the bifurcating families have been observed in various bar models.

The family $x = \dot{x} = 0$ is unstable between the two bifurcating orbits, while one of the bifurcating families is stable and the other unstable. This is true if ϵ is absolutely small. However, in some cases of large perturbations the bifurcating families may have opposite orientations close to their origin (e.g., one opens toward larger r_c and the other toward smaller r_c). In such a case both families may be stable (FIG. 6a) or unstable.

If the perturbation is very strong, the "central" family has one more point of

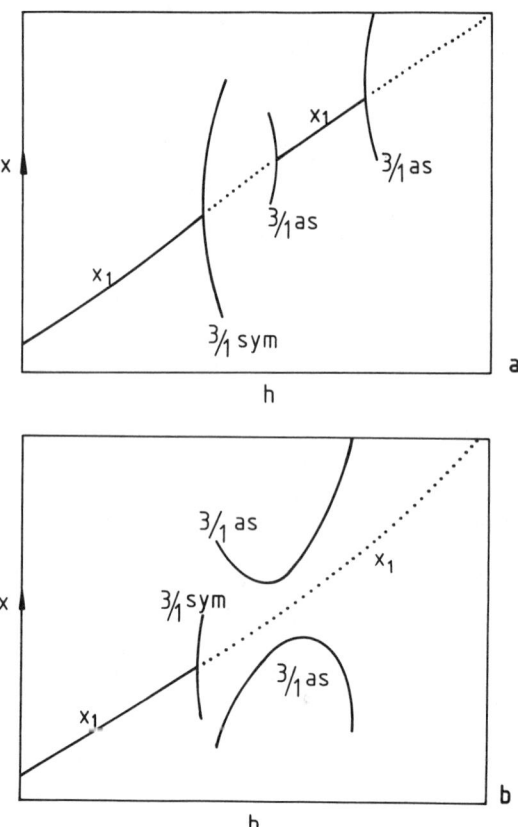

FIGURE 6. The characteristics of the 3/1 families (symmetric and asymmetric with respect to the x axis, perpendicular to the bar) for strong bars (schematically). *Solid lines* indicate stable orbits, and *dotted lines* unstable orbits. Case (**b**) represents a stronger bar than case (**a**).

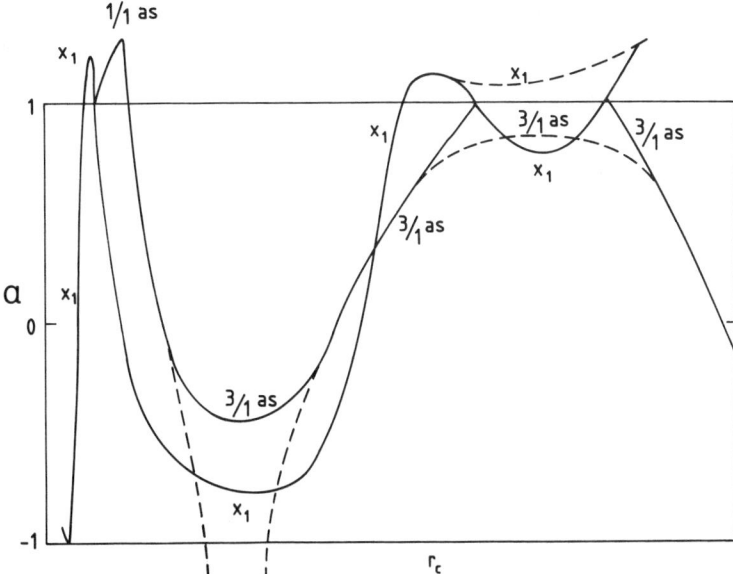

FIGURE 7. Stability curves of the families x_1 and 3/1 (asymmetric). The Hénon stability parameter, a, is given as a function of the radius of the circular orbit with the same energy (schematically). The orbits with $a > 1$ or $a < -1$ are unstable. The *dashed lines* represent a stronger bar than the *solid lines*.

bifurcation, beyond which it is unstable (FIGS. 6a and 7). At this point of bifurcation starts a new 3/1 family of asymmetric orbits. For an even larger perturbation the two asymmetric bifurcating families join and separate from the central family. Then the central family is unstable over a large interval beyond the symmetric 3/1 bifurcation, while the asymmetric bifurcating family is stable (FIGS. 6b and 7, dashed lines).

The orbits of the asymmetric 3/1 family have a figure-eight form (FIG. 8). The 1/1 asymmetric orbits are also figure eight, but they are much closer to the center.

Orbits of this type were also found by Pfenniger (reference 16, fig. 5b), but the loops are only barely seen in a few cases. On the other hand, in our models[13] the figure-eight loops are very clear (FIG. 8). Furthermore, in our models the 3/1 asymmetric orbits are stable over a large interval of r_c (or of energy, h).

Many figure-eight orbits were found empirically in a collapsed N-body system by Miller and Smith.[18] Such orbits are mostly stable and extend up to two thirds of corotation (end of the bar). This distance is of the same order as the 3/1 resonance, while the 1/1 resonance is much closer to the center[12] (about 1/10 of corotation).

In some bar models the characteristics of the 3/1 families extend to large values of r_c, and then return to smaller r_c, until they reach the characteristic of the retrograde family x_4 (figs. 1a, 3, and 5 of Contopoulos[19]). If the bar is sufficiently strong, one may have also one or two 1/1 families close to the center (one asymmetric family, or one asymmetric and one symmetric; Martinet[20]). These families bifurcate from the central family x_1 at two points, and their characteristics form closed bubbles, unrelated to the 3/1 families.

However, in other strong models a different scenario is common. Namely the 3/1 families are joined with the 1/1 families that start much closer to the center (FIG. 7). This phenomenon was observed already by Pfenniger (reference 16, fig. 6). We found the same behavior in many models of the form of equation 23, that we have explored recently. In most models we found that the asymmetric 1/1 orbits bifurcating from the "central" family are unstable for small r_c, become stable over a small interval of r_c, again unstable over another interval, and then stable for a large interval of values of r_c (FIG. 7, dashed line). In particular when the "central" family is mostly unstable the 3/1 asymmetric family is the main stable family of periodic orbits.

Therefore, the 3/1 resonant orbits seem to play an important role in the dynamics of barred galaxies. This is consistent with the results of Miller and Smith,[18] who find

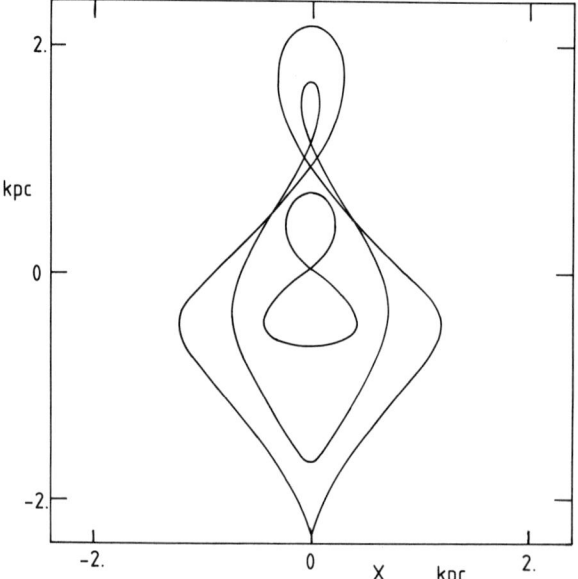

FIGURE 8. Figure-eight orbits that belong to the 3/1 asymmetric family.

that in a galactic model produced by an N-body collapse, about 25–50% of the orbits are of this type.

THREE DEGREES OF FREEDOM

Resonant systems can also be found in three degrees of freedom. Such systems can be written in action-angle variables in the form

$$H = \omega_1 I_1 + \omega_2 I_2 + \omega_3 I_3 + aI_1^2 + 2bI_1I_2 + cI_2^2 + 2dI_1I_3 + 2eI_2I_3 + fI_3^2$$

$$+ \Sigma \, \epsilon_{k_1 k_2 k_3} \genfrac{}{}{0pt}{}{\cos}{\sin} (k\theta_1 + k_2\theta_2 + k_3\theta_3). \tag{34}$$

If there is resonance relation between ω_1, ω_2, ω_3 (accurate, or approximate) of the form

$$k_1\omega_1 + k_2\omega_2 + k_3\omega_3 \simeq 0 \tag{35}$$

for given k_1, k_2, k_3, then the most important trigonometric terms depend on a combination $(k_1\theta_1 + k_2\theta_2 + k_3\theta_3)$. If we omit all other terms, we have an integrable system, with two more integrals besides H, namely

$$J_2 = k_1 I_2 - k_2 I_1 \qquad J_3 = k_1 I_3 - k_3 I_1. \tag{36}$$

If there are two independent resonance relations between ω_1, ω_2, ω_3 of the form

$$k_1\omega_1 + k_2\omega_2 + k_3\omega_3 \simeq 0,$$

$$m_1\omega_1 + m_2\omega_3 + m_3\omega_3 \simeq 0,$$

and we have only the corresponding combinations of angles in equation 34, then there is one extra integral of motion, of the form

$$J_3 = (k_2 m_3 - k_3 m_2) I_1 + (k_3 m_1 - k_1 m_3) I_2 + (k_1 m_2 - k_2 m_1) I_3. \tag{37}$$

This is a partially integrable case, which can be separated into a system of two degrees of freedom, plus an independent motion along a third dimension.

An interesting property of partially separable 3-dimensional systems is that they do not have any complex instability along any family of periodic orbits. Complex instability means that the eigenvalues of the variational equations corresponding to a periodic orbit are complex, but not on the unit circle.[21,22] Such complex instability appears when two pairs of stable eigenvalues collide on the unit circle (as one parameter varies) and then one pair moves outward and the other inward, but not on the real axis. This transition is not possible if the system is partially separable, because the two pairs of eigenvalues move independently of each other.

The nonexistence of complex instability in a particular 3-dimensional system may lead one to suspect that this system is partially separable. This was the case of the system

$$H = \tfrac{1}{2}(\dot{x}^2 + \dot{y}^2 + \dot{z}^2 + x^2 + y^2 + z^2) - \epsilon x z^2 - \eta y z^2 \tag{38}$$

studied by Contopoulos and Barbanis,[23] which does not show any complex instability. It was later realized that this system has a second integral

$$\Phi = \tfrac{1}{2} [(\eta \dot{x} - \epsilon \dot{y})^2 + (\eta x - \epsilon y)^2], \tag{39}$$

but no third integral of motion.

A systematic exploration of partially integrable systems of three degrees of freedom, like the study of Hietarinta,[24] would be particularly useful as regards applications in galactic dynamics.

SUMMARY

We have studied generic systems of two degrees of freedom in action-angle variables. Hamiltonians that contain only one combination of angles $(m\theta_1 - n\theta_2)$ are integrable. Such Hamiltonians describe the resonance phenomena that appear when the radial and rotational frequencies ω_1 and ω_2 have a ratio close to n/m. In bisymmetric galactic models the potential contains only trigonometric terms of the form $\cos n\theta$, where n is even. If the ratio n/m is even, we find a gap along the characteristic of the "central" family of periodic orbits (the one reduced to circles in the unperturbed case), while if n/m is odd, the "central" family has an unstable segment. As examples we studied the resonances 4/1 and 3/1. If the perturbation is of the form $\epsilon(\cos 2\theta + \alpha_4 \cos 4\theta)$, the orientation of the gap depends on α_4 and ϵ. In the case of the 3/1 resonance one resonant family, bifurcating from the "central" family, contains figure-eight orbits. In some models these orbits are stable over a large energy interval. Thus we can explain the observed abundance of figure-eight orbits in the collapsed N-body systems of Miller and Smith. The preceding results can be extended to systems of three degrees of freedom. Such systems may have one or two resonant combinations of angles. In the first case the system is integrable, while in the second case it is partially integrable. Such systems do not have any complex unstable periodic orbits.

REFERENCES

1. KUZMIN, G. G. 1953, 1956. Tartu. Astron. Obs. Teated, Nos. 1, 2, 3.
2. HORI, G. 1962. Publ. Astron. Soc. Japan, **14**: 353.
3. VAN DE, HULST, H.C. 1962. Bull. Astron. Inst. Neth. **16**: 235.
4. DE ZEEUW, P. J. & D. LYNDEN-BELL. 1985. Mon. Not. R. Astron. Soc. **215**: 713.
5. DE ZEEUW, P. J. 1985. Mon. Not. R. Astron. Soc. **216**: 273.
6. DE ZEEUW, P. J., C. HUNTER & M. SCHWARZSCHILD. 1987. Astrophys. J. **317**: 607.
7. MAYER, F. & L. MARTINET. 1973. Astron. Astrophys. **27**: 199.
8. CONTOPOULOS, G. 1975. Astrophys. J. **201**: 566.
9. CONTOPOULOS, G. 1983. Physica **8D**: 142.
10. CONTOPOULOS, G. & TH. PAPAYANNOPOULOS. 1980. Astron. Astrophys. **92**: 33.
11. TEUBEN, P. J. & R. H. SANDERS. 1985. Mon. Not. R. Astron. Soc. **212**: 257.
12. PAPAYANNOPOULOS, TH. & M. PETROU. 1983. Astron. Astrophys. **119**: 21.
13. CONTOPOULOS, G., S. T. GOTTESMAN, J. H. HUNTER & M. N. ENGLAND. 1988. Preprint.
14. PAPAYANNOPOULOS, TH. 1988. In preparation.
15. ATHANASSOULA, E., O. BIENAYME, L. MARTINET & D. PFENNIGER. 1983. Astron. Astrophys. **127**: 349.
16. PFENNIGER, D. 1984. Astron. Astrophys. **134**: 373.
17. CONTOPOULOS, G. 1988. Astron. Astrophys. In press.
18. MILLER, R. H. & B. F. SMITH. 1979. Astrophys. J. **227**: 785.
19. CONTOPOULOS, G. 1983. Celest. Mech. **31**: 193.
20. MARTINET, L. 1984. Astron. Astrophys. **132**: 381.
21. BROUCKE, R. 1969. Am. Inst. Aeronaut. Astronaut. J. **7**: 1003.
22. CONTOPOULOS, G. & P. MAGNENAT. 1985. Celest. Mech. **37**: 387.
23. CONTOPOULOS, G. & B. BARBANIS. 1985. Astron. Astrophys. **153**: 44.
24. HIETARINTA, J. 1988. New integrable systems. This volume.

Integrable Models for Galaxies[a]

TIM DE ZEEUW

Institute for Advanced Study
Princeton, New Jersey 08540

THE SELF-CONSISTENT PROBLEM

The structure and dynamics of a collisionless stellar system such as a galaxy are determined completely by the phase–space distribution function $f(\mathbf{r}, \mathbf{v}, t)$, which gives the distribution of the stars in the system over position $\mathbf{r}$ and velocity $\mathbf{v}$ as a function of the time t. The distribution function satisfies the collisionless Boltzmann equation

$$\frac{\partial f}{\partial t} + \mathbf{v} \cdot \nabla f - \nabla V \cdot \frac{\partial f}{\partial \mathbf{v}} = 0, \qquad (1)$$

where V is the potential in which the stars move, and $\dot{\mathbf{v}} = -\nabla V$ by Newton's equations. Since it is generally assumed that galaxies are in a steady state, f does not depend on time explicitly, and $\partial f / \partial t = 0$.

The density $\rho(\mathbf{r})$ of the system is the integral of f over the velocities

$$\rho(\mathbf{r}) = \iiint f(\mathbf{r}, \mathbf{v}) \, d^3\mathbf{v}. \qquad (2)$$

In a self-gravitating system V is the gravitational potential of the density itself, and is related to ρ via Poisson's equation

$$\nabla^2 V = 4\pi G \rho. \qquad (3)$$

In order to construct a dynamical model for a self-gravitating stellar system in equilibrium, the preceding equations have to be solved simultaneously. A solution corresponds to a dynamical model only if $f \geq 0$. The problem of finding f for a stellar system (in equilibrium) is the fundamental problem of stellar dynamics.[1] It is often referred to as the self-consistent problem.

Jeans' theorem states that f depends on the phase–space coordinates $(\mathbf{r}, \mathbf{v})$ only through the isolating integrals of motion admitted by the potential.[2,3] Any such function automatically satisfies equation 1. Hence one is left with equations 2 and 3. Since the number of isolating integrals is generally limited to at most three, Jeans' theorem allows a reduction of the number of variables in the self-consistent problem. Evidently, this requires knowledge of all the isolating integrals admitted by the potentials of interest.

A spherical potential generally possesses four isolating integrals. In addition to the energy E, the angular momentum vector $\mathbf{L} = (L_x, L_y, L_z)$ is conserved. As a result, the general distribution function of a spherical galaxy is of the form $f(E, \mathbf{L})$. If one requires that f is spherical in all its properties, it can depend only on the magnitude of L

---

[a]This work was supported in part by National Science Foundation Grant PHY 8620266, and in part by an RCA fellowship.

but not on its direction, so that $f = f(E, L^2)$. Such models have anisotropic velocity distributions. For $f = f(E)$ the distribution is isotropic in velocity space.

Eddington[4] showed in 1915 that, for a given $\rho(r)$, it is always possible to invert equation 2 explicitly in order to obtain a unique $f(E)$. If $\rho(r)$ falls off with radius sufficiently rapidly, this $f(E)$ is nowhere negative, and it represents the unique isotropic solution. Many more anisotropic solutions $f(E, L^2)$ exist. They can sometimes be found by analytic inversion techniques, but they are usually constructed by assumption of a special functional form for f, or by numerical techniques.[5,6]

AXISYMMETRIC GALAXIES

An axisymmetric potential always admits two exact isolating integrals, the energy E and the component of the angular momentum vector that is parallel to the symmetry axis, L_z, say. Lynden-Bell[7] generalized Eddington's inversion formula, and showed that, for a given axisymmetric density $\rho(\varpi, z)$, one can always find a unique $f(E, L_z^2)$ that is consistent with it. In such models the velocity dispersion $\langle v_\varpi^2 \rangle$ in the equatorial plane is everywhere equal to the velocity dispersion $\langle v_z^2 \rangle$ in the z direction. The method for the actual calculation of f is not easy to apply in practice,[5] and only a few models have been constructed in this manner. Usually one assumes a functional form for f, and then solves the self-consistent problem by series expansions or by numerical means.[8]

The standard example of an axisymmetric system is our own galaxy. However, the work of Kapteyn, Jeans, and Oort in the early decades of this century showed that $\langle v_\varpi^2 \rangle \neq \langle v_z^2 \rangle$ in the solar neighborhood.[9] It was concluded that the distribution function of the Galaxy must depend on a third argument, that is, there must be a third isolating integral of motion, I_3. Although special potentials that admit a third integral were considered by various authors,[10-13] it was established only through the analytic work of Contopoulos[14] and the numerical work of Ollongren[15] that in realistic galactic potentials, indeed, most stellar orbits have an effective third integral.[16]

In nearly spherical potentials the third integral is related closely to L^2, which is an integral in the spherical limit.[17] This fact has been used in order to derive approximate distribution functions $f = f(E, L_z^2, I_3)$ for models with density distributions that are appropriate for galaxies.[8,18] Since a solution of the self-consistent problem with $f = f(E, L_z^2)$ is guaranteed by Lynden-Bell's algorithm, it is evident that many solutions with $f = f(E, L_z^2, I_3)$ exist.

Most orbits in realistic axisymmetric potentials are tubes around the symmetry axis. The remainder is generally made up of a host of minor orbit families and irregular orbits. The latter do not have a third integral of motion. Binney has argued that Jeans' theorem is not strictly valid for potentials that support irregular orbits.[19,20] Hence, no true equilibrium solutions may exist for such systems. In many cases of interest, however, the fraction of irregular orbits is small.[21] For practical purposes one may probably still use Jeans' theorem for these systems, and construct approximate equilibrium models.

TRIAXIAL GALAXIES

Although a two-integral model was known to be inadequate for the Galaxy, it was nevertheless thought for a long time that elliptical galaxies have a simpler dynamical structure, and are oblate axisymmetric systems with $f = f(E, L_z)$ and an isotropic velocity distribution. This premise generated most of the work described in the previous section. However, spectroscopic observations performed in the mid-1970s indicated that large elliptical galaxies rotate much slower than expected.[22,23] Binney[24] showed that the then available kinematic data required velocity distributions that are anisotropic, that is, $f = f(E, L_z, I_3)$. He further concluded that it would be more natural to assume that these galaxies are in fact triaxial and have a figure that is at most slowly rotating. Many observational lines of evidence support this hypothesis.[25,26]

Triaxial potentials generally admit only one exact isolating integral, the orbital energy E. Although there are three planes of reflection symmetry, there are no symmetry axes and no component of the angular momentum vector is conserved. Numerical orbit calculations show that in slowly rotating triaxial potentials relevant for elliptical galaxies, most stellar orbits belong to a few major families and possess two effective integrals, I_2 and I_3, in addition to the energy.[27-29]

With the exception of a few special solutions,[30-32] triaxial self-consistent models have so far either been *simulated* in N-body experiments,[33-35] or they have been *constructed* by numerical techniques that sidestep our ignorance concerning two of the three integrals of motion. Following Schwarzschild, one reconstructs an assumed density distribution from the densities of individual orbits in the corresponding potential, subject to the constraint that the occupation numbers must be nonnegative. This has been done by linear programming, by application of Lucy's iterative technique, by means of a nonnegative least squares method, or by a maximum entropy approach.[27-29,36-40]

STÄCKEL MODELS

There are many potentials for which the Hamilton–Jacobi equation separates, so that they admit three exact isolating integrals of motion. Special cases were already discussed about 150 years ago.[41-43] The potentials have been named after Stäckel, who was the first to classify them systematically, around the turn of the century.[44-47] Eddington rediscovered the same potentials in his study of dynamical models obeying the ellipsoidal hypothesis.[48] Since then, the Stäckel potentials have appeared in numerous investigations in mathematical physics.[49-51]

For the most general case, the Hamilton-Jacobi equation separates in ellipsoidal coordinates (λ, μ, ν), and the potential V_s must have the form[52]

$$V_s = -\frac{F(\lambda)}{(\lambda - \mu)(\lambda - \nu)} - \frac{F(\mu)}{(\mu - \nu)(\mu - \lambda)} - \frac{F(\nu)}{(\nu - \lambda)(\nu - \mu)}, \quad (4)$$

where $F(\tau)$ is an arbitrary function ($\tau = \lambda, \mu, \nu$). For a given ellipsoidal coordinate system, and a given smooth function $F(\tau)$, a potential of the form of equation 4 corresponds to a mass model that is everywhere smooth and has a nonrotating triaxial figure.[53-55] Since the (λ, μ, ν) coordinates are determined by specification of two parameters—the positions of the foci on the z axis—each choice of $F(\tau)$ will lead to a two-parameter family of triaxial Stäckel models, the parameters being the central axis ratios of the density distribution.

Every orbit in a Stäckel model has three exact isolating integrals of motion. E, I_2, and I_3, which are known explicitly, and which are quadratic in the velocities. The integrals I_2 and I_3 are related to the angular momentum integrals of the axisymmetric and spherical limits.[56] The orbit can be considered as the sum of three motions, one in each ellipsoidal coordinate. The stars are thus constrained—by the integrals of motion—to move between coordinate surfaces. Thus, all possible orbital shapes can be found by inspection of the ellipsoidal coordinate system in which the motion separates. It turns out that all centrally concentrated triaxial mass models of this kind contain four families of orbits: boxes, short-axis tubes, and two families of long-axis tubes.[37,55] These are exactly the four major orbit families that occur in Schwarzschild's[27] nonrotating triaxial model.

The prototypical triaxial Stäckel model is the perfect ellipsoid. It has a density distribution given by

$$\rho = \rho(m^2) = \frac{\rho_0}{(1 + m^2)^2} \qquad m^2 = \frac{x^2}{a^2} + \frac{y^2}{b^2} + \frac{z^2}{c^2}, \qquad (5)$$

with $a \geq b \geq c$. The oblate case, $a = b$, was discovered by Kuzmin in his classic study of separable models of the Galaxy.[12] He also obtained the general case, and showed that it has four major orbit families.[53] The perfect ellipsoid was rediscovered by de Zeeuw and Lynden-Bell, who showed that it is the only inhomogeneous triaxial mass model with a Stäckel potential in which the density is stratified exactly on similar concentric ellipsoids.[56]

The Stäckel models have some remarkable properties. Kuzmin[12,53] showed that in a Stäckel model the density $\rho(x, y, z)$ at a general point is related to that on the short axis, $\rho(0, 0, z)$, by a very simple formula, and that $\rho(x, y, z) \geq 0$ if and only if $\rho(0, 0, z) \geq 0$. This makes it possible to choose a short-axis density profile, and the values of the central-axis ratios of the model, and find the complete mass model that has a Stäckel potential and this density profile, by a one-dimensional quadrature. A second integration gives this potential explicitly.[55] Hence, Poisson's equation can be integrated in closed form for the whole class of Stäckel models.

De Zeeuw, Peletier, and Franx constructed many different mass models and delineated their general properties.[57] Models with a singular density in the center only do not exist. The density cannot fall off more rapidly than r^{-4} as $r \to \infty$, except on the short axis. Models in which ρ falls off less rapidly than r^{-4} become spherical as $r \to \infty$. The only models that have surfaces of constant density that approach a finite flattening at large radii are those with $\rho \sim r^{-4}$. The projected surface density of a triaxial Stäckel model generally has isophotes that show a change in ellipticity with radius, but no twisting.[58]

NONUNIQUENESS

The Stäckel potentials support no irregular orbits, and therefore Jeans' theorem is valid, so that $f = f(E, I_2, I_3)$. Although the three integrals of motion are known explicitly, direct inversion of equation 2 in order to obtain $f(E, I_2, I_3)$ from a given triaxial density $\rho(x, y, z)$ is a rather intimidating task. However, the individual orbit densities in a Stäckel model are known in analytic form, and hence are evaluated easily. This makes it straightforward to construct self-consistent models by means of Schwarzschild's method, while avoiding laborious numerical orbit integrations. This is true not only for the triaxial models, but also for the various limiting cases with more symmetry, and a simpler orbital structure.[8]

For the oblate and prolate axisymmetric limits, I_2 reduces to $\frac{1}{2}L_z^2$, and it is evident that many different distribution functions $f = f(E, I_2, I_3)$ will be consistent with a given density. Bishop[59] has constructed a number of such distribution functions for the perfect oblate spheroid, that is, the model defined in equation 5 with $b = a$. He first specified one-dimensional continua of orbits, by choosing the dependence of f on one integral, and then solved for the remaining dependence by means of an algebraic technique developed by Vandervoort.[60]

The triaxial case was investigated by Statler.[37] He considered 21 different perfect ellipsoidal mass models. For each of these he constructed one solution by Lucy's method, thus establishing existence, and 15-20 distinct solutions by means of linear programming. The properties of the models vary smoothly with the axis ratios. The fundamental result of Statler's study is that many different dynamical models can be constructed with the same density distribution, as already suggested by Schwarzschild's original models.[27,28]

By analogy with the spherical and axisymmetric case one might have expected that for a triaxial Stäckel model the density, which depends on three variables, uniquely determines a distribution function f that depends on all three integrals. One could further argue that even in case of a unique distribution function, the numerical approaches might produce a number of different solutions due to discretization effects. However, one can also make a case for nonuniqueness. Each of the four (major) orbit families in the Stäckel potential covers all of configuration space. As a result, there may be ample opportunity to exchange orbits, so that different combinations of orbits make up the same density distribution. This argument holds for nonseparable triaxial potentials as well.

It is important to establish whether or not a given triaxial density distribution can have many different f's consistent with it. If the solutions are unique, then by observing kinematic properties of elliptical galaxies we may learn only relatively little about their formation. If the dynamical solutions are nonunique, more photometric and kinematic observations are required to determine the intrinsic structure of elliptical galaxies, but these will provide substantial insight in the processes that were responsible for the formation of these galaxies.

Stars on tube orbits have a definite sense of rotation around either the long axis of the model or around the short axis. Since the Stäckel models have nonrotating figures, clockwise and counterclockwise motion may occur in the same tube orbit. The present discussion ignores the nonuniqueness that is introduced by simply reversing the velocity vector of an arbitrary fraction of stars in each tube orbit. Nonuniqueness means that combinations of orbits with truly different shapes produce the same density distribution.

In order to settle this issue, de Zeeuw, Hunter, and Schwarzschild investigated the analogous two-dimensional problem.[61] For $c = 0$ the models defined in equation 5 reduce to elliptic disks. These contain flat box orbits, and flat short-axis tubes. Teuben used Schwarzschild's method to construct equilibrium models for nine different axis ratios, both with minimum and with maximum possible streaming, and discussed the kinematic properties.[62] De Zeeuw, Hunter, and Schwarzschild used analytic methods to prove rigorously that exact equilibrium models exist, and showed that there is a freedom of a function of two variables in the construction of distribution functions f that are consistent with the same ρ. The fundamental reason for the nonuniqueness of the dynamical solutions is the existence of more than one major orbit family, so that orbits can be exchanged while keeping the density fixed and $f \geq 0$. These results strongly indicate that the self-consistent solutions for triaxial Stäckel models are nonunique as well, and that the freedom is one of a function of three variables. The same conclusion was drawn by Dejonghe, who gave a formal solution of the self-consistent problem for this case.[63]

EXACT SOLUTIONS

Since many properties of the Stäckel models can be given in analytic form, it is natural to ask whether it is possible to give exact distribution functions $f(E, I_2, I_3)$. For an arbitrary potential of the form of equation 4 this is not to be expected, since every spherical potential is of Stäckel form, and only a few analytic spherical models exist. However, it is possible to obtain some special solutions.

Thin Orbit Models

For any oblate Stäckel potential the special equilibrium model that contains infinitesimally thin short-axis tubes only (i.e., tubes without radial epicyclic motion) can be found by simple inversion of a one-dimensional Abel equation.[64] The corresponding distribution function can be given explicitly, and the kinematic properties can be calculated by one-dimensional quadratures. If all stars are assumed to have the same sense of rotation around the symmetry axis, the models have the maximum amount of streaming that is consistent with the given density distribution. Numerical solutions of this kind have been given by Bishop.[65] The structure of the similar prolate models is more complicated, due to the presence of two types of long axis tubes.[66]

Triaxial models with maximum streaming, containing thin tube orbits of the three families as well as boxes, can also be constructed in nearly explicit form, by generalizing the analysis of the perfect elliptic disk mentioned in section 5. They are discussed in more detail by Hunter elsewhere in this volume.

Fricke's Method

Fricke[67] showed that in a general axisymmetric potential $V(\varpi, z)$, a distribution function f of the form

$$f(E, L_z^2) = \sum_{l,m} a_{lm} E^l L_z^{2m}, \qquad (6)$$

corresponds to a density ρ given by

$$\rho(\varpi, z) = 2\pi \sum_{l,m} a_{lm} \frac{\Gamma(l+1)\Gamma(m+\tfrac{1}{2})}{\Gamma(l+m+\tfrac{1}{2})} 2^m \varpi^{2m} V^{l+m+3/2}. \qquad (7)$$

Thus, if a given density $\rho(\varpi, z)$ in a potential $V(\varpi, z)$ can be expressed as $\rho(\varpi, V)$, and can be expanded in the form of equation 7, then the unique distribution function $f(E, L_z^2)$ that is consistent with it follows immediately from equation 6. A drawback of this approach is that it is not always clear whether the series in equation 6 converges. Still, a number of the available two-integral axisymmetric models have been constructed in this manner.[8]

For models with a Stäckel potential it is possible to extend Fricke's method to three integral distribution functions.[68] If one writes f as

$$f(E, I_2, I_3) = \sum_{l,m,n} a_{lmn} E^l I_2^m (I_2 + I_3)^n, \tag{8}$$

where l and m may take real values, and n is an integer, then the corresponding density distribution in a given Stäckel potential is of the form

$$\rho(\varpi, z) = \sum_{l,m,n} a_{lmn} \rho_{lmn}(\varpi, z), \tag{9}$$

where the individual components ρ_{lmn} can be written down explicitly. The advantage of taking powers of $I_2 + I_3$ rather than of I_3 lies in the fact that $I_2 + I_3$ reduces to L^2 in the spherical limit, while the required calculations are no more complicated. Since $I_2 = \frac{1}{2} \cdot L_z^2$, equations 8 and 9 reduce to Fricke's result for $n = 0$.

Expansion of a given density in a series of the form of equation 9 gives f by comparison of coefficients. Each choice of such a series for ρ will result in a different distribution function. Although the convergence of the resulting series is not guaranteed, this method in principle produces a distribution function that depends on all three integrals of motion. In practice, there is unfortunately little hope that the required series expansion for ρ can be found in closed form, due to the cumbersome form of the individual components ρ_{lmn}. However, the series expansion may be done by numerical means, by fitting the components ρ_{lmn} to a given density distribution. The same method can be applied to triaxial systems also, and should give smooth solutions.

For axisymmetric models, the generalized Fricke method may be combined with one of the available two-integral inversion methods[5] in order to produce exact three-integral models.[68] If $f(E, L_z^2)$ is the two-integral distribution function for the density distribution ρ of the whole model, then the distribution function

$$f(E, L_z^2, I_3) = f(E, L_z^2) + \sum_{l,m,n} a_{lmn} [E^l I_2^m (I_2 + I_3)^n - f_{lmn}(E, L_z^2)], \tag{10}$$

where $f_{lmn}(E, L_z^2)$ is the two-integral distribution function for an individual component ρ_{lmn}, is consistent with ρ. Thus, by choosing different combinations of components ρ_{lmn} one obtains different three-integral distribution functions. In order to obtain fully analytic solutions, the following three requirements must be met:

1. The density ρ can be written explicitly as $\rho(\varpi, V)$. This is a necessary, although not sufficient, condition for a successful application of the two-integral inversion algorithm[5,7] for the calculation of $f(E, L_z^2)$.
2. The mass model has a Stäckel potential, so that the calculation of the three-integral components ρ_{lmn} is valid.

3. The densities ρ_{lmn} can be written explicitly as $\rho_{lmn}(\varpi, V)$, where V is the potential of the whole model, so that $f_{lmn}(E, L_z^2)$ can be calculated.

To date, only one axisymmetric (ρ, V)-pair is known for which ρ can be written explicitly as $\rho(\varpi, V)$, and V is of Stäckel form.[8] This pair describes a one-parameter family of axisymmetric mass models that contains the Kuzmin[69] disk and Hénon's[70] spherical isochrone as special cases. It was introduced by Kuzmin,[12] and subsequently studied by Kuzmin and Kutuzov.[71] Dejonghe and de Zeeuw[68] have shown recently that for this family of models the third requirement mentioned in the preceding is satisfied also. These authors constructed a number of exact distribution functions of the form of equation 10, and discussed the kinematic properties of the resulting dynamical models.

OTHER POTENTIALS. ROTATION

The potentials of elliptical galaxies, and also that of our own galaxy, can be approximated fairly accurately by the special potentials of Stäckel form.[54,72] Numerical orbit calculations have shown that the orbital structure in the Stäckel models is typical for moderately flattened triaxial systems with a nonrotating figure, and also for mildly flattened axisymmetric systems. Although in general additional orbit families occur, as well as some stochastic orbits, these invariably occupy a small fraction of phase–space. Thus, most orbits will have three integrals of motion. This is confirmed by perturbation theory.[73,74] However, the orbital structure in strongly flattened systems, or in cases where figure rotation is important, can differ substantially from that in a Stäckel model. As a result, the dynamical models discussed in the previous section, and the ones still under construction, should be very useful in the delineation of the detailed internal dynamical structure of elliptical galaxies, and also of the halos of spiral galaxies, but should not be used for barred galaxies.

Whereas the Stäckel potentials are the only potentials in nonrotating coordinates for which the Hamilton–Jacobi equation separates,[50] there might be special cases where the equations of motion admit extra integrals, although the motion is not separable. A number of such integrable cases are known, especially in two dimensions, but to date mostly for potentials that are not relevant for galaxies.[75] It will be interesting to establish whether realistic integrable triaxial potentials $V = V(x^2, y^2, z^2)$ exist. In this respect it should be noted that scale-free triaxial potentials for which numerical orbit calculations have been performed seem to possess a remarkably small fraction of stochastic orbits.[74,76] This suggests that there might be scale-free potentials that are exactly integrable.

To date, little work has been done on the determination of the integrals of motion admitted by rotating potentials. Except for a special case found by Vandervoort,[77] the only known rotating potential for which the equations of motion separate is the three-dimensional rotating harmonic oscillator, which is the gravitational potential of a homogeneous ellipsoid.[30] It is very important to establish whether or not more integrable cases exist.[78] Here one should concentrate the search for extra integrals on triaxial potentials, that is, potentials of the form $V(x^2, y^2, z^2)$.

SUMMARY

Dynamical models with a Stäckel potential, for which the Hamilton–Jacobi equation separates in ellipsoidal coordinates, give considerable insight into the structure of axisymmetric as well as nonrotating triaxial galaxies. Triaxial stellar systems with appreciable figure rotation require much further study.

ACKNOWLEDGMENTS

It is a pleasure to thank Dr. G. Contopoulos for his invitation to participate in this workshop. The work on the generalization of Fricke's method has been done in collaboration with Herwig Dejonghe. He, Marijn Franx, and Daniel Pfenniger commented on the manuscript.

REFERENCES

1. CHANDRASEKHAR, S. 1942. Principles of Stellar Dynamics. Univ. of Chicago Press. Chicago.
2. JEANS, J. H. 1915. Mon. Not. R. Astron. Soc. **76**: 71–84.
3. LYNDEN-BELL, D. 1962. Mon. Not. R. Astron. Soc. **124**: 1–9.
4. EDDINGTON, A. S. 1915. Mon. Not. R. Astron. Soc. **76**: 572–585.
5. DEJONGHE, H. 1986. Phys. Rep. **133**: 218–313.
6. BINNEY, J. J. & S. D. TREMAINE. 1987. Galactic Dynamics. Princeton Univ. Press. Princeton, N.J.
7. LYNDEN-BELL, D. 1962. Mon. Not. R. Astron. Soc. **123**: 447–458.
8. DE ZEEUW, P. T. 1987. In Proc. IAU Symp. Structure and Dynamics of Elliptical Galaxies. P. T. de Zeeuw, Ed.: 271–290. Reidel. Dordrecht.
9. OORT, J. H. 1965. In Galactic Structure, Stars and Stellar Systems, Vol. 5. A. Blaauw and M. Schmidt, Eds.: 455–511. Univ. of Chicago Press. Chicago.
10. VAN ALBADA, G. B. 1952. Proc. Kon. Ned. Akad. Wet., Series B. **55**: 620–627.
11. LINDBLAD, B. 1959. Handbuch der Physik, Vol. 53: 21–99. Springer-Verlag. Berlin/New York.
12. KUZMIN, G. G. 1956. Astron. Zh. **33**: 27–45.
13. HORI, G. 1962. Publ. Astron. Soc. Japan. **14**: 353–373.
14. CONTOPOULOS, G. 1960. Z. Astrophys. **49**: 273–291.
15. OLLONGREN, A. 1962. Bull. Astron. Inst. Neth. **16**: 241–295.
16. MARTINET, L. & F. MAYER. 1975. Astron. Astrophys. **44**: 45–57.
17. SAAF, A. 1968. Astrophys. J. **154**: 483–498.
18. PETROU, M. 1983. Mon. Not. R. Astron. Soc. **202**: 1195–1208, 1209–1220.
19. BINNEY, J. J. 1982. Mon. Not. R. Astron. Soc. **201**: 15–19.
20. PFENNIGER, D. 1986. Astron. Astrophys. **165**: 74–83.
21. GOODMAN, J. & M. SCHWARZSCHILD. 1981. Astrophys. J. **245**: 1087–1093.
22. BERTOLA, F. & M. CAPACCIOLI. 1975. Astrophys. J. **200**: 439–445.
23. ILLINGWORTH, G. D. 1977. Astrophys. J. Lett. **218**: L43–L47.
24. BINNEY, J. J. 1978. Mon. Not. R. Astron. Soc. **183**: 501–514.
25. DAVIES, R. L. 1987. In Proc. IAU Symp. Structure and Dynamics of Elliptical Galaxies. P. T. de Zeeuw, Ed.: 63–78. Reidel. Dordrecht.
26. SCHECHTER, P. L. 1987. In Proc. IAU Symp. Structure and Dynamics of Elliptical Galaxies. P. T. de Zeeuw, Ed.: 217–228. Reidel. Dordrecht.
27. SCHWARZSCHILD, M. 1979. Astrophys. J. **232**: 236–247.
28. SCHWARZSCHILD, M. 1982. Astrophys. J. **263**: 599–612.

29. LEVISON, H. & D. O. RICHSTONE. 1987. Astrophys. J. **314:** 476–492.
30. FREEMAN, K. C. 1966. Mon. Not. R. Astron. Soc. **133:** 47–62; **134:** 1–14, 15–23.
31. HUNTER, C. 1974. Mon. Not. R. Astron. Soc. **166:** 633–648.
32. VANDERVOORT, P. O. 1980. Astrophys. J. **240:** 478–487; **241:** 316–333.
33. AARSETH, S. J. & J. J. BINNEY. 1978. Mon. Not. R. Astron. Soc. **185:** 227–243.
34. WILKINSON, A. & R. A. JAMES. 1982. Mon. Not. R. Astron. Soc. **199:** 171–196.
35. VAN ALBADA, T. S. 1987. In Proc. IAU Symp. Structure and Dynamics of Elliptical Galaxies. P. T. de Zeeuw, Ed.: 291–299. Reidel. Dordrecht.
36. VIETRI, M. 1986. Astrophys. J. **306:** 48–63.
37. STATLER, T. S. 1987. Astrophys. J. **321:** 113–152.
38. LUCY, L. B. 1974. Astron. J. **79:** 745–754.
39. PFENNIGER, D. 1984. Astron. Astrophys. **141:** 171–188.
40. TREMAINE, S. D. & D. O. RICHSTONE. 1987. Preprint.
41. LIOUVILLE, J. 1846. J. Math. **11:** 345–378; **14:** 257–299.
42. BERTRAND, J. 1852. J. Math. **17:** 121–174.
43. JACOBI, C. G. J. 1866. Vorlesungen über Dynamik. Reimer. Berlin.
44. STÄCKEL, P. 1890. Math. Ann. **35:** 91–103.
45. STÄCKEL, P. 1893. Math. Ann. **42:** 537–563.
46. LEVI-CIVITA, T. 1904. Math. Ann. **59:** 383–397.
47. DALL'ACQUA, F. A. 1908. Math. Ann. **66:** 398–415.
48. EDDINGTON, A. S. 1915. Mon. Not. R. Astron. Soc. **76:** 37–60.
49. EISENHART, L. P. 1934. Ann. of Math. **35:** 284–305.
50. LYNDEN-BELL, D. 1962. Mon. Not. R. Astron. Soc. **124:** 95–123.
51. MILLER, JR., W. 1977. Symmetry and Separation of Variables. Addison-Wesley. Reading, Mass.
52. WEINACHT, J. 1924. Math. Ann. **91:** 279–299.
53. KUZMIN, G. G. 1973. In Dynamics of Galaxies and Clusters. T. B. Omarov, Ed.: 71–75. Akademiya Nauk Kazakhskoj SSR, Alma Ata. (Transl. in Proc. IAU Symp. Structure and Dynamics of Elliptical Galaxies. P. T. de Zeeuw, Ed.: 553–557. Reidel. Dordrecht.)
54. DE ZEEUW, P. T. 1985. Mon. Not. R. Astron. Soc. **216:** 273–333.
55. DE ZEEUW, P. T. 1985. Mon. Not. R. Astron. Soc. **216:** 599–612.
56. DE ZEEUW, P. T. & D. LYNDEN-BELL. 1985. Mon. Not. R. Astron. Soc. **215:** 713–730.
57. DE ZEEUW, P. T., R. F. PELETIER & M. FRANX. 1986. Mon. Not. R. Astron. Soc. **221:** 1001–1022.
58. FRANX, M. 1988. Mon. Not. R. Astron. Soc. **231:** 285–308.
59. BISHOP, J. 1986. Astrophys. J. **305:** 14–27.
60. VANDERVOORT, P. O. 1984. Astrophys. J. **287:** 475–486.
61. DE ZEEUW, P. T., C. HUNTER & M. SCHWARZSCHILD. 1987. Astrophys. J. **317:** 607–636.
62. TEUBEN, P. 1987. Mon. Not. R. Astron. Soc. **227:** 815–841.
63. DEJONGHE, H. 1987. Submitted for publication in SIAM J. Math. Anal.
64. DE ZEEUW, P. T. 1988. In preparation.
65. BISHOP, J. 1987. Astrophys. J. **322:** 618–631.
66. HUNTER, C., P. T. DE ZEEUW, CH. PARK & M. SCHWARZSCHILD. 1988. In preparation.
67. FRICKE, W. 1952. Astron. Nachr. **280:** 193–216.
68. DEJONGHE, H. & P. T. DE ZEEUW. 1988b. Astrophys. J. In press.
69. KUZMIN, G. G. 1953. Tartu. Astron. Obs. Teated. **1.**
70. HÉNON, M. 1959. Ann. Astrophys. **22:** 126–139.
71. KUZMIN, G. G. & S. A. KUTUZOV. 1962. Bull. Abastumani Astrophys. Obs. **27:** 82–88.
72. DEJONGHE, H. & P. T. DE ZEEUW. 1988a. Astrophys. J. **329.** In press.
73. DE ZEEUW, P. T. 1985. Mon. Not. R. Astron. Soc. **215:** 731–760.
74. GERHARD, O. E. 1985. Astron. Astrophys. **151:** 279–296.
75. HIETARINTA, J. 1987. Phys. Rep. **147:** 87–154.
76. RICHSTONE, D. O. 1982. Astrophys. J. **252:** 496–507.
77. VANDERVOORT, P. O. 1979. Astrophys. J. **232:** 91–105.
78. SCHWARZSCHILD, M. 1987. In Chaotic Phenomena in Astrophysics. Ann. N.Y. Acad. Sci. **479:** 16–21.

Integrable Galactic Models[a]

C. HUNTER

Department of Mathematics
Florida State University
Tallahassee, Florida 32306-3027

INTRODUCTION

This paper gives a brief summary of recent progress on the problem of finding a self-consistent stellar dynamic model for a specified potential of Stäckel form.[1] Such a model is described by a density distribution f in the six-dimensional phase space of position **r** and velocity **v**. By Jeans'[2] theorem, f can depend only on the three isolating integrals that motion in a Stäckel potential possesses. Self-consistency requires that

$$\int f d^3\mathbf{v} = \rho(\mathbf{r}), \tag{1}$$

where $\rho(\mathbf{r})$ is the spatial density of mass needed to produce the Stäckel potential. A particular example of such a density that has been used frequently in computational work is that of the so-called perfect ellipsoid[3]

$$\rho = \frac{M}{\pi^2 abc[1 + x^2/a^2 + y^2/b^2 + z^2/c^2]^2}. \tag{2}$$

Here M is the total mass, while a, b, and c, with ordering $a \geq b \geq c$, are the semiaxes of the ellipsoidal isodensity surfaces.

The mathematical problem to be discussed, therefore, is that of solving equation 1 with a known ρ for an unknown f. Equation 1 is thus an integral equation of the first kind. As is well known, such equations usually cannot be solved explicitly. We have attempted, in the work to be described, to work analytically as far as possible, and, to this end, have made much use of thin tube orbits. However, part of the work has had to be done computationally, with some specific choice of density, such as that of equation 2.

Dynamics in a Stäckel potential is simple because the motion in each coordinate proceeds independently and is bounded by certain coordinate surfaces.[1] Consequently, the geometry of these surfaces has a profound influence on the construction of self-consistent models. The coordinate surfaces λ = const. are ellipsoids, the surfaces μ = const. are hyperboloids of one sheet that are threaded by the x axis, while the surfaces ν = const. are hyperboloids of two sheets, the x–y plane lying between the sheets.[1] Values of the coordinates are restricted to the following abutting ranges:

$$\lambda \geq a^2 = -\alpha$$
$$a^2 \geq \mu \geq b^2 = -\beta$$
$$b^2 \geq \nu \geq c^2 = -\gamma. \tag{3}$$

---

[a]This work was supported in part by National Science Foundation Grants DMS-8420624 and DMS-8701228.

The negative quantities α, β, and γ, to which the semiaxes introduced earlier are related, are the fundamental parameters of the coordinate system.

The extreme degenerate cases of the coordinate surfaces are significant in what follows. One is $\lambda = a^2$, a flattened ellipsoid that occupies that part of the plane $x = 0$ interior to the *focal ellipse* $y^2/(a^2 - b^2) + z^2/(a^2 - c^2) = 1$. The remainder of the plane $x = 0$ is given by $\mu = a^2$, and is a collapsed hyperboloid of one sheet. The other extreme value of $\mu = b^2$ is another degenerate hyperboloid of one sheet, but one that has been collapsed to the simple-connected part of the plane $y = 0$ interior to the *focal hyperbola* $z^2/(b^2 - c^2) - x^2/(a^2 - b^2) = 1$. The outer two parts of this plane are given by $\nu = b^2$, a degenerate hyperboloid of two sheets.

Not only is motion in a Stäckel potential restricted to lie between certain coordinate surfaces, but only four different combinations of restrictions, in addition to those of inequalities 3, can occur.[1] The four allowed combinations correspond to four different classes of orbits and are:

Box orbits: $\lambda \leq \lambda_0$, $\quad \mu \leq \mu_0$, $\quad \nu \leq \nu_0$ $\qquad$ (4a)

Inner long (x-) axis tubes: $\lambda \leq \lambda_0$, $\quad \mu_1 \leq \mu \leq \mu_2$ $\qquad$ (4b)

Outer long axis tubes: $\lambda_1 \leq \lambda \leq \lambda_2$, $\quad \mu_0 \leq \mu$ $\qquad$ (4c)

Short (z-) axis tubes: $\lambda_1 \leq \lambda \leq \lambda_2$, $\quad \nu \leq \nu_0$. $\qquad$ (4d)

The terms long and short here refer to the axes of the mass distribution (cf. equation 2) about which the orbits rotate. Graphic illustrations of the outlines of the four orbital types are given in figures 2a–2d of Statler.[4]

Each of the three subscripted quantities in each of cases 4 is a constant of the motion, and it is convenient to use these constants, the so-called turning-point coordinates, as the isolating integrals of motion.[5] Each orbital type then has its own distribution function and its own density interrelated by equation 1. Moreover, apart from some exceptions on degenerate coordinate surfaces, every point in coordinate space is accessible to orbits of each type. Because it is the sum of the separate densities only that must match that associated with the Stäckel potential, the possibility of a variety of self-consistent solutions arises.

Before discussing the general ellipsoidal case, we shall first discuss three simpler limiting cases. These are, in increasing order of complexity:

(1) The oblate spheroid, which is obtained in the limit $b \to a$ and allows only short z-axis tubes.
(2) The prolate spheroid, which is obtained in the limit $c \to b$. It allows both kinds of tubes about the long x axis.
(3) The elliptic disk, which is obtained when the limits $c \to 0$ and $\rho \to \infty$ of the density 2 are taken in such a way that there remains surface density $\Sigma = M/2\pi ab[1 + x^2/a^2 + y^2/b^2]^{3/2}$ on the plane $z = 0$. Both box orbits and short-axis tubes in the plane are available to reproduce this surface density.

THE ELLIPTIC DISK[6]

Though this is not the simplest of the special cases, it does provide the closest paradigm to the full three-dimensional problem. The coordinate surfaces in the plane

$z = 0$ ($\nu = c^2$) are confocal ellipses λ = const. and hyperbolae μ = const. Box orbits are restricted to regions of the form $a \leq \lambda \leq \lambda_0$, $b^2 \leq \mu \leq \mu_0$, such as that shaded in FIGURE 1. The tube orbits, on the other hand, are unrestricted in μ, but are confined to a region of the form $\lambda_1 \leq \lambda \leq \lambda^2$ between two of the confocal ellipses. If we denote by Σ_b and Σ_t the respective surface densities of the two types of orbits, then we require that

$$\Sigma_t + \Sigma_b = \Sigma. \tag{5}$$

From the orbital dynamics, we obtain two separate integral equations of the general form of equation 1, which are

$$\frac{1}{2} \int_{a^2}^{\lambda} d\lambda_1 \int_{\lambda}^{\infty} d\lambda_2 \frac{F_t(\lambda_1, \lambda_2)}{\sqrt{D(\lambda; \lambda_1, \lambda_2)} \sqrt{-D(\mu; \lambda_1, \lambda_2)}} = \Sigma_t(\lambda, \mu) \tag{6a}$$

$$\frac{1}{2} \int_{\mu}^{a^2} d\mu_0 \int_{\lambda}^{\infty} d\lambda_0 \frac{F_b(\mu_0, \lambda_0)}{\sqrt{D(\lambda; \mu_0, \lambda_0)} \sqrt{-D(\mu; \mu_0, \lambda_0)}} = \Sigma_b(\lambda, \mu). \tag{6b}$$

These equations are written entirely in turning-point coordinates. The distribution functions in these coordinates, F_b and F_t for boxes and tubes, respectively, differ from the phase space density f by Jacobian factors. The denominators of equations 6 contain functions $D(\tau; \tau_1, \tau_2)$, which are of the form

$$D(\tau; \tau_1, \tau_2) = (\tau_2 - \tau)(\tau - \tau_1) R(\tau, \tau_1, \tau_2), \tag{7}$$

where the function R is a divided difference of a basic function that appears in the Stäckel potential. The properties of R that are essential to the present discussion are that it is a positive, everywhere analytic, and symmetric function of its arguments.

The problem now is that of finding physically acceptable nonnegative solutions of equations 6 for both F_t and F_b. There are two further constraints in addition to equation 5. As equation 6b shows, a nonsingular box distribution function F_b cannot provide any of the density on the degenerate hyperbola $\mu = a^2$, so that

$$\Sigma_b(\lambda, a^2) = 0. \tag{8}$$

This is because $\mu = a^2$ is that part of the y axis outside the foci, that is, a degenerate hyperbola. It is reached only by the outer edges of boxes of extreme shape with $\mu_0 = a^2$. Similarly, a nonsingular tube distribution F_t cannot provide any density on $\lambda = a^2$, that is, the line $S_1 S_2$ between the foci of FIGURE 1. The constraint 8 plus the fact that the integral equation 6b for F_b is of the Volterra type, while equation 6a is of a more complicated type, suggest the following method of solution. We first select some distribution function F_t that produces the required density on the outer parts of the y axis, and for which $\Sigma_t(\lambda, a^2) = \Sigma(\lambda, a^2)$. This selection can be carried out analytically when the tube orbits are all infinitesimally thin closed ellipses. From the F_t that is obtained thereby, we can compute the tube density $\Sigma_t(\lambda, \mu)$ for the whole x–y plane, and thus the right-hand size $\Sigma_b = \Sigma - \Sigma_t$ for equation 6b. The final step is then the solution of this Volterra integral equation for F_b. Although this final step has to be carried out numerically to confirm that $F_b \geq 0$ throughout, standard methods from integral equation theory can be used to establish the existence of a unique solution for F_b that is bounded everywhere and continuous everywhere except at the foci $\lambda = \mu = a^2$. Away from the foci, the integral equation 6b is of the generalized Abel type because

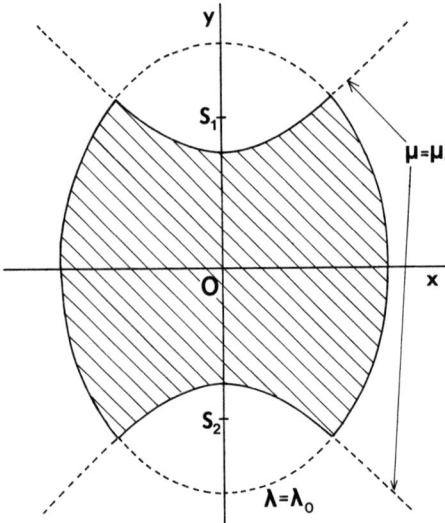

FIGURE 1. The region shaded is that filled by a typical box orbit, and is bounded by a coordinate surface of each type. All the coordinate surfaces are conics with foci S_1, S_2, at $x = 0$, $y = \pm(a^2 - b^2)^{1/2}$.

one of the linear factors of each D vanishes at the lower limits of integration. However, both linear factors of each D can vanish as the foci are approached, and a more complicated analysis is needed there. Its outcome is that F_b is discontinuous but bounded at the foci where curves of constant, though finite, F_b meet.

The solution that has just been described is not unique. To show formally that a second solution can be constructed, we first select some set of nonclosed tube orbits, and then show that the remaining portion of the density Σ can be reproduced with closed elliptical tubes and boxes in the same manner as before. Evidently, a considerable variety of solutions is possible now that there are two distinct classes of orbits to provide the density.

THREE-DIMENSIONAL FIGURES

In the full three-dimensional problem, we have four integral equations, one for each class of orbits, rather than the two equations 6a and 6b. That for the short-axis tubes, for instance is,

$$\frac{1}{2\sqrt{2}} \int_\nu^{b^2} d\nu_0 \int_{a^2}^\lambda d\lambda_1 \int_\lambda^\infty d\lambda_2$$

$$\cdot \frac{F_s(\nu_0, \lambda_1, \lambda_2)}{\sqrt{B(\lambda; \nu_0, \lambda_1, \lambda_2)} \sqrt{-B(\mu; \nu_0, \lambda_1, \lambda_2)} \sqrt{B(\nu; \nu_0, \lambda_1, \lambda_2)}} = \rho_s(\lambda, \mu, \nu), \quad (9)$$

where

$$B(\tau; \tau_1, \tau_2, \tau_3) = (\tau - \tau_1)(\tau - \tau_2)(\tau_3 - \tau) R(\tau, \tau_1, \tau_2, \tau_3), \quad (10)$$

and R is another symmetric and everywhere positive analytic divided difference function.

Oblate Spheroids

Short-axis tubes are the only possible orbits in the special case of this $b = a$ limit, so that equation 9 is the only integral equation to be solved now. Its solution is highly nonunique because the distribution function F_s remains a function of three integrals, whereas the density reduces to the two variable function $\rho_s(\lambda, a^2, \nu)$, now that $\mu = b^2 = a^2$ everywhere.

The reduced form of equation 9 can be solved explicitly if, as in the disk problems, we restrict attention to infinitesimally thin tubes.[7,8] These are now tubes that lie always on the part of some ellipsoid $\lambda = $ const. that lies between the two sheets of the hyperboloid $\nu = $ const. (see FIG. 2). The appropriate distribution function is then of the form

$$F_s(\nu_0, \lambda_1, \lambda_2) = \mathcal{F}_s(\nu_0, \lambda_1)\delta(\lambda_2 - \lambda_1), \tag{11}$$

and equation 9 reduces to

$$\int_\nu^{a^2} \frac{\tilde{F}_s(\nu_0, \lambda) d\nu_0}{\sqrt{(\nu_0 - \nu)R(\nu, \nu_0, \lambda, \lambda)}} = \frac{2\sqrt{2}(\lambda - a^2)(\lambda - \nu)\rho_s(\lambda, a^2, \nu)}{\pi}, \tag{12}$$

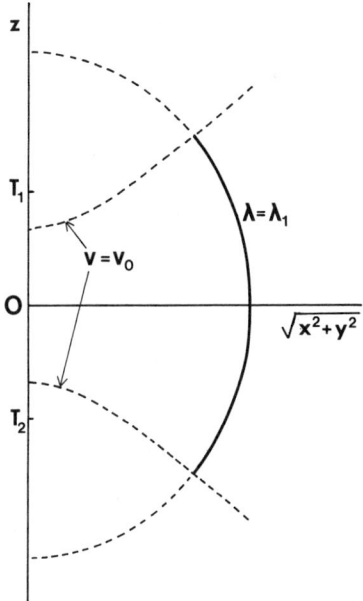

FIGURE 2. A plane through the axis of symmetry of an oblate spheroid. A thin tube fills a surface such as that obtained by rotating the thick elliptic segment shown about the axis of symmetry. All the coordinate surfaces in this plane have foci T_1, T_2 at $x = y = 0$, $z = \pm(a^2 - c^2)^{1/2}$.

after the unknown distribution function has been redefined as

$$\tilde{F}_s(\nu_0, \lambda) = \frac{\mathcal{F}_s(\nu_0, \lambda)}{\sqrt{(\lambda - \nu_0)(a^2 - \nu_0)R(\nu_0, \lambda, \lambda, \lambda)R(\nu_0, a^2, \lambda, \lambda)}}. \tag{13}$$

Explicit solution of equation 12 is possible because R is a divided difference, so that the term in the square root in the denominator can be rewritten as the difference between some function of ν and the same function of ν_0. Equation 12 therefore is precisely an Abel integral equation when that function of ν_0 is made to be the variable of integration. The solution of equation 12 is straightforward. An interesting feature of its solution[7] is that the distribution function $\mathcal{F}_s(\nu_0, \lambda)$ is now well-behaved even at the degenerate hyperboloid $\nu_0 = a^2$. This degenerate surface has now shrunk to that part of the z axis outside the foci. But although the density here must be provided by the short-axis tubes with $\nu_0 = a^2$, which reach this axis at an extremity, geometrical convergence to a circle of zero radius at this extremity allows a finite distribution function to provide the finite density needed.

Prolate Spheroids[9]

The long x axis is now the axis of symmetry in this case in which $\nu = b^2 = c^2$ everywhere. Inner and outer long-axis tubes are possible, and there is more redundancy than in the oblate case because both have three-variable distribution functions. Both distribution functions satisfy equations similar to equation 9. A key part of this problem, like that of the elliptic disk, is a splitting

$$\rho(\lambda, \mu, b^2) = \rho_i(\lambda, \mu, b^2) + \rho_0(\lambda, \mu, b^2) \tag{14}$$

of the density into the parts provided by the two kinds of tubes. This splitting is also subject to constraints on the degenerate coordinate surfaces $\lambda = a^2$ and $\mu = a^2$ in the equatorial $x = 0$ plane. These constraints are

$$\rho_i(\lambda, a^2, b^2) = 0 \qquad \rho_0(a^2, \mu, b^2) = 0, \tag{15}$$

because a nonsingular distribution of inner long-axis tubes gives zero density on the part of the plane $x = 0$ outside the focal circle, whereas outer long-axis tubes give zero density on the inner part of this plane. FIGURE 3 illustrates the geometry in the plane of symmetry when, again, thin tubes only are used. Equations of the Abel type and analogous to equation 12 are then obtained, and can be solved explicitly for both kinds of tubes. However, further restrictions must be added to equations 15 if distribution functions that are everywhere finite in action variables are to be obtained.

Ellipsoids of General Form[10]

Not only are all four classes of orbits now possible, but they are all needed if a density such as that of equation 2 is to be produced by a nonsingular distribution of orbits. The reason for this is again the inability of finite distribution functions to provide nonzero densities on certain degenerate coordinate surfaces. On the collapsed two-sheeted hyperboloid $\nu = b^2$, both ρ_s (cf. equation 9) and $\rho_b = 0$. Both kinds of long-axis tubes can provide the density on this surface. Finding a combination of them that does so is exactly equivalent to solving the prolate spheroid problem (cf. equation 14). The density on the collapsed one-sheeted hyperboloid $\mu = a^2$, on the other hand, can be provided by either outer long-axis or short-axis tubes. When the distribution of outer long-axis tubes has been decided upon at the previous step, finding a distribution of short-axis tubes that reproduces the remaining density at $\mu = a^2$ is the same as solving the oblate spheroid problem. Then, once distributions of all three kinds of tube

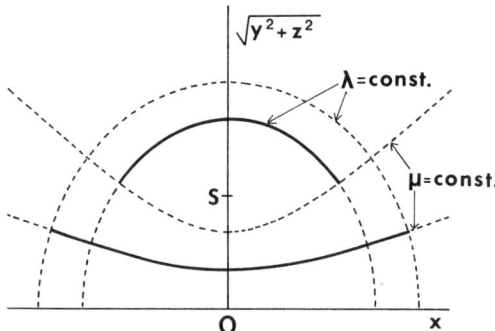

FIGURE 3. A plane through the axis of symmetry of a prolate spheroid. Thin outer/inner long-axis tubes fill the surfaces obtained when elliptic/hyperbolic segments, such as the thick ones shown, are rotated about the x axis of symmetry. The coordinate surfaces in this plane all share the focus S at $x = 0$, $(y^2 + z^2)^{1/2} = (a^2 - b^2)^{1/2}$. When rotated, S traces out the focal circle.

orbits have been found, $\rho_b = \rho - \rho_0 - \rho_i - \rho_s$ is known and it remains only to solve the Volterra equation

$$\frac{1}{2\sqrt{2}} \int_\nu^{b^2} d\nu_0 \int_\mu^{a^2} d\mu_0 \int_\lambda^\infty d\lambda_0$$

$$\cdot \frac{F_b(\nu_0, \mu_0, \lambda_0)}{\sqrt{B(\lambda; \nu_0, \mu_0, \lambda_0)} \sqrt{-B(\mu; \nu_0, \mu_0, \lambda_0)} \sqrt{B(\nu; \nu_0, \mu_0, \lambda_0)}} = \rho_b(\lambda, \mu, \nu), \quad (16)$$

for F_b. This equation is a three-dimensional version of equation 6b. It is likewise of the generalized Abel type, and the same analytical methods as before can be applied to it. The only complications that occur do so on the *focal ellipse* $\lambda = \mu = a^2$ and the *focal hyperbola* $\mu = \nu = b^2$. These two curves are two of the outer edges of the rectangular box of ellipsoidal coordinate space in which equation 16 must be solved. The complications are that four linear factors in the square-root denominator vanish at these two edges, where the situation is like that at the focus of the elliptic disk. The result is that F_b is now discontinuous along these two edges at which surfaces of constant, though finite, values of F_b meet. Consequently, all the essential features of the full three-dimensional problem have been met, and resolved, in the three special limiting cases that were discussed earlier.

REFERENCES

1. DE ZEEUW, T. 1985. Mon. Not. R. Astron. Soc. **216:** 273–334.
2. JEANS, J. H. 1915. Mon. Not. R. Astron. Soc. **76:** 70–84.
3. DE ZEEUW, P. T. & D. LYNDEN-BELL. 1985. Mon. Not. R. Astron. Soc. **215:** 713–730.
4. STATLER, T. S. 1987. Astrophys. J. **321:** 113–152.
5. VANDERVOORT, P. O. 1984. Astrophys. J. **287:** 475–486.
6. DE ZEEUW, P. T., C. HUNTER & M. SCHWARZSCHILD. 1987. Astrophys. J. **317:** 607–636.
7. DE ZEEUW, T. In preparation.
8. BISHOP, J. L. 1987. Astrophys. J. **322:** 618–631.
9. HUNTER, C., T. DE ZEEUW, C. PARK, & M. SCHWARZSCHILD. In preparation.
10. HUNTER, C. & T. DE ZEEUW. In preparation.

New Integrable Systems[a]

JARMO HIETARINTA

Department of Physical Sciences
University of Turku
Turku, Finland

GENERAL

In this talk we are going to discuss the integrability of Hamiltonian systems, mainly of two degrees of freedom. Furthermore the motion is assumed to take place in flat space, that is, the Hamiltonian has the form

$$H = \tfrac{1}{2}(p_x^2 + p_y^2) + Ap_x + Bp_y + V, \tag{1}$$

where A, B, and V are functions of x and y. For part of our discussion we take $A = B = 0$, but we include also some nonzero cases, which are used, for example, in the description of motion in electromagnetic fields, the coriolis effect, and the Fokker-Planck systems.

Since we will discuss "integrability," we must first define the term. Several definitions are in use and a nonnegligible amount of confusion has been created when this same term has been used with different definitions.

Definition: The Hamiltonian H in equation 1 is called (Liouville) *integrable* if there exists another function $I = I(p_x, p_y, x, y)$ such that

$$\{H, I\} = 0, \tag{2}$$

where the braces stand for the Poisson bracket. The function I must also have the following properties: (1) it is functionally independent of H, (2) it exist globally and is single valued, and (3) it is a complex analytic function of its variables. This definition generalizes to N-dimensional systems: In general one needs N functions I_k ($I_1 = H$, say) such that they are in involution, that is, $\{I_k, I_m\} = 0$, $\forall\ 1 \leq k, m \leq N$.

The function I just given is called a second invariant or constant of motion or isolating integral. It is assumed to be time independent.

Condition (3) is the one that has caused most confusion. There are some rigorous methods to show nonintegrability, (Painlevé[1] and Ziglin[2]), but to apply them it is essential to consider motion in the complex time plane. Thus these methods can show only the nonexistence of *complex analytic* second invariants, while at the same time one can often by other means show that the system has a real analytic constant of motion (e.g., by taking the initial values). So when numerical methods indicate that a system might be integrable they can only suggest the existence of a real analytic invariant, but numerical results usually yield no information about the complex

---

[a]This work was supported by the Academy of Finland.

analytic structure of the invariant. Complex analytic (or even algebraic) invariants seem to be necessary when we want to solve the system explicitly, because only then do we have at our disposal sufficiently powerful mathematical methods (see, e.g., reference 3).

We will now go through the known integrable systems of the type given in equation 1, giving several examples. In most cases we will just give the integrable potential and the order of the invariant; the precise form of the second invariant can be found from the given references. For further material, see reference 4 and references therein.

LINEAR AND QUADRATIC INVARIANTS (A = B = 0)

The search for integrable Hamiltonians has a long history that goes back at least to Bertrand, although the first systematic studies were probably made by Darboux.[5] His results were included in Whittaker's influential book,[6] and the search for polynomial (in momenta) invariants came to be called the Whittaker program.

In this kind of search problem one must take into account the various invariances to simplify the situation as much as possible. For example, if $A = B = 0$ in equation 1, then H is invariant under $p_x \to -p_x$, $p_y \to -p_y$, and therefore, since the Poisson bracket respects this invariance, we may assume that $I(-p_x, -p_y, x, y) = \pm I(p_x, p_y, x, y)$,[4] and treat the odd and even cases separately. Similarly, one can and should use the basic linear transformations (translation, scaling, and rotation) to simplify the system.[4]

Let us start with the ansatz that the second invariant is linear in momenta. The Hamiltonians with such an invariant are easily obtained, and one finds that a linear I solves equation 2 if

$$I = a(xp_y - yp_x) + bp_x + cp_y \tag{3}$$

$$V = f(\tfrac{1}{2}a(x^2 + y^2) - cx + by), \tag{4}$$

where a, b, and c are arbitrary constants and f an arbitrary function.

Darboux was the first to study the quadratic invariants,[5] but his results were incomplete (he assumed that one of the free constants was nonzero). More complete results were given later by several authors independently.[7-11]

It is easy to show that a sufficiently general ansatz for a quadratic I is

$$I = A(xp_y - yp_x)^2 + (xp_y - yp_x)(ap_x + bp_y)$$
$$+ c_0 p_x^2 + c_1 p_x p_y + c_2 p_y^2 + D(x, y), \tag{5}$$

where A, a, b, and c_i are constants. The final integrable results can be transformed into the following five distinct cases:[12]

(1) Generic Case:
$$V = [f(u) + g(v)]/(u + v), \tag{6a}$$

$$u = -[A(x^2 + y^2) + ax - by - c_0 + c_2] + (\partial_x R^2 + \partial_y R^2)^{1/2}, \tag{6b}$$

$$v = [A(x^2 + y^2) + ax - by - c_0 + c_2] + (\partial_x R^2 + \partial_y R^2)^{1/2}, \tag{6c}$$

$$R = A(x^3/3 - xy^2) + \tfrac{1}{2}a(x^2 - y^2) + bxy + (c_0 - c_2)x + c_1 y. \tag{6d}$$

(2) $A \neq 0, a = b = c_i = 0$:
$$V = f(r) + g(\theta)r^{-2}. \tag{7}$$
(3) $A = 0, a = -1, b = \mp i, c_1 = \pm i(c_0 - c_2)$:
$$V = f(x \pm iy)/r + g(x \pm iy). \tag{8}$$
(4) $A = a = b = c_1 = c_2 = 0, c_0 = 1$:
$$V = f(x) + g(y). \tag{9}$$
(5) $A = a = b = c_2 = 0, c_0 = 1, c_1 = \pm i$:
$$V = r^2 f(x \pm iy) + g(x \pm iy). \tag{10}$$

As special cases, (1) contains potentials separable in elliptic and parabolic coordinates.

HIGHER ORDER INVARIANTS ($A = B = 0$)

As we saw the Whittaker problem has been solved completely for quadratic invariants, but only partial results are known for higher order invariants.[4] Systematic attempts with restricted ansatze have been made to find cubic[13] and quartic[14] invariants, and some isolated results also exist with sixth-order invariants. Multidimensional systems that have been studied with the Lax-pair approach[15] have also produced systems whose two-dimensional versions have higher order invariants.

Here we will give integrable potentials of some particular types. For further results, see reference 4.

(1) The Hénon–Heiles potential
$$V = \mu/3\, y^3 + x^2 y \tag{11}$$
has been studied for cubic and quadratic invariants by several authors. For $\mu = 1, 6$, the Hamiltonian has a quadratic invariant; for $\mu = 16$ a quartic invariant;[16,17] no other case is known to be integrable. Several additional terms can be added to this potential without destroying integrability.

(2) The Holt potential[13]
$$V = (\mu/3)\, y^{4/3} + x^2 y^{-2/3} \tag{12}$$
is closely associated with the cubic potential given previously: It is known to be integrable only for $\mu = 1, 6, 16$, and in fact there exists a (noncanonical) transformation to connect it to the cubic model.[18] The second invariants for these potentials are of order 2, 3 and 6, respectively.

(3) The quartic potential
$$V = ax^4 + bx^2 y^2 + cy^4 \tag{13}$$

has also been studied often. It has a quadratic invariant if $a:b:c = a:0:c$ or $1:2:1$ or $1:6:1$ or $1:12:16$ or $16:12:1$. It is also integrable for $a:b:c = 1:6:8$ or $8:6:1$, but then the invariant is the quartic.[19,20]

(4) Toda-type potentials, for example, the "two-particle open-end" Toda lattice

$$V = e^{x+\alpha y} + e^{x-\alpha y}, \tag{14}$$

have been of great interest due to their connection with Lie algebras.[21-23] Using Ziglin's analysis, one can show[24] that equation 14 can have a complex analytic second invariant only if $\alpha = n(n-1)/2$ for some integer n. For $n = 1, 2,$ and 3 the invariant is known and has order 1, 2, and 3, respectively.

POLYNOMIAL INVARIANTS WITH NONZERO A AND/OR B

The models with nonzero A and/or B have been studied much less. The equations for polynomial invariants are more complicated because there is no reflection symmetry to simplify the situation. There is, however, a new symmetry that can be used, namely gauge invariance: $p_x \to p_x + \Phi_x, p_y \to p_y + \Phi_y$, where $\Phi = \Phi(x, y)$ is an arbitrary function. The gauge invariant quantities that characterize the system are

$$U = A_y - B_x, \qquad W = V - \tfrac{1}{2}(A^2 + B^2). \tag{15}$$

For a linear ansatz a complete solution can be obtained fairly easily. One finds the integrable pair of potentials

$$U = g(\tfrac{1}{2}a(x^2 + y^2) - cx + by)$$
$$W = f(\tfrac{1}{2}a(x^2 + y^2) - cx + by), \tag{16}$$

where g and f are arbitrary functions.

Even for a quadratic invariant the problem is still open. Some of the particular cases that have been studied include restricting the invariant to $I = p_x^2 +$ lower order terms,[25] or the potential U to a constant (with totally even invariant).[26]

RATIONAL INVARIANTS

The invariants just given were all polynomials in the momenta, but there is no fundamental reason why we could not have more complicated invariants. It is perhaps just more difficult to find them, and therefore they are not so familiar to us.

For a rational invariant we make the ansatz $I = R/S$, where R and S are polynomials (relatively prime), and compute the Poisson bracket with the Hamiltonian:

$$\{H, I\} = [\{H, R\}S - \{H, S\}R]S^{-2}. \tag{17}$$

If the expression in square brackets is to vanish, we must have

$$\{H, R\} = GR, \qquad \{H, S\} = GS, \tag{18}$$

where G is some polynomial. (If $G = 0$, we can take R as an invariant by itself.) Any nontrivial solution of equations 18 provides us with a rational invariant.

If we assume that the leading (highest order in the momenta) parts of R and S commute with H, we find that G is independent of momenta, and that the potential U in definition 15 must be nonzero. The case where R and S are furthermore linear in momenta has been solved in reference 4. As an example we can have

$$R = xp_y - yp_x + h(x, y), \qquad S = p_y + k(x, y), \tag{19}$$

with

$$A = x, \qquad B = \tfrac{1}{2}(x^2 - y^2)/y, \qquad C = 0, \qquad U = -x/y,$$
$$W = -(x^2 + y^2)y^{-2}/8, \qquad h = -2xy, \qquad k = 0, \tag{20}$$

or [27]

$$A = 0, \quad B = -x/y, \quad C = 0, \quad U = 1/y, \quad W = -\tfrac{1}{2}(x/y)^2, \quad h = y, \quad k = 0. \tag{21}$$

TRANSCENDENTAL INVARIANTS

Transcendental invariants can be obtained by generalizing the procedure given above: We assume that the invariant is a more general function of two polynomials, $I = I(R, S)$, say. This is a generalization of the Whittaker program in the sense that we now have two polynomials rather than one.

For simplicity assume that R and S are two different, but of same order, polynomials, whose highest order parts commute with the kinetic term of the Hamiltonian. Then

$$\{H, R\} = G(lR + mS + n)$$
$$\{H, S\} = G(tR + uS + v), \tag{22}$$

for some constants l, m, n, t, u, v, and function $G(x, y)$. The Poisson bracket now becomes

$$\{H, I\} = [I_R(lR + mS + n) + I_S(tR + uS + v)]G. \tag{23}$$

Thus I is an invariant, if the term in the square brackets vanishes, which gives us an equation for I once we have a solution of equations 22. Transcendental invariants are obtained, for example, when $l + u \neq 0$, $lv - tn \neq 0$, $lu - tm = 0$. A solution to equation 23 is then given by $I = [(lR + mS)(l + u) + ln + mv] \exp[-(lS - tR)(l + u)/lv - tn)]$.

Several results with R and S linear in the momenta were obtained in reference 4, for example:

(1) $l = 1, m = n = u = 0, \; t, v \neq 0$: $A = -e^{-(x+iy)} + vy, \quad B = ivy, \; C = 0,$
$$U = te^{-(x+iy)} + v, \qquad W = -\tfrac{1}{2}e^{-2(x+iy)} + vye^{-(x+iy)},$$
$$R = p_x, \qquad S = p_y + v(x + iy), \qquad I = R \exp[(tR - S)/v]. \tag{24}$$

(2) $l = 1, m = n = u = t = 0, \quad v \neq 0: \quad A = vy/x + dx, \quad B = 0, \quad C = 0,$

$U = v/x, \quad W = \tfrac{1}{2}A^2, \quad R = p_x + 2dx, \quad S = p_y + v\log x, \quad I = R\exp[-S/v].$ (25)

Finally, to show that a really simple Hamiltonian can have a complicated invariant we would like to present the following Hamiltonian[27]

$$H = \tfrac{1}{2}(p_x^2 + p_y^2) + x/y,$$ (26)

and invariants

$$I_2 = \tfrac{1}{2}y[p_y W_+(\tfrac{1}{2}E, p_x) + 2W'_+(\tfrac{1}{2}E, p_x)]^2,$$

$$I_3 = [p_y W_-(\tfrac{1}{2}E, p_x) + 2W'_-(\tfrac{1}{2}E, p_x)]/[p_y W_+(\tfrac{1}{2}E, p_x) + 2W'_+(\tfrac{1}{2}E, p_x)],$$ (27)

where the W's are the parabolic cylinder functions and $E = H$. Since there are two extra invariants, this system is superintegrable, and in fact $[I_2, I_3] = 1$.

INTEGRABILITY FOR ZERO ENERGY

For the invariants just given, the Poisson bracket between H and I vanished identically, but it is, of course, possible that some expressions are constants of motion only for some special sets of initial values. One such situation is met when

$$\{H, I\} = GH$$ (28)

for some function G (which is nonsingular when $H = 0$). A function I with property 28 is therefore invariant only if the motion is such that $H = 0$, which may happen in some practical situations.

The assumption $H = 0$ can be interpreted as a constraint

$$0 = \tfrac{1}{2}(p_x^2 + p_y^2) + V(x, y).$$ (29)

In principle we can then solve p_y, say, and eliminate it from the invariant and from the equation $\{H, I\} = 0$. Thus we will get a smaller set of equations. However, this will introduce square roots, so it is more practical to first make a complex rotation to

$$0 = p_z p_w + W(z, w)$$ (30)

and then eliminate p_w, for example.

Another advantage of working at zero energy is the existence of a large class of transformations that keep the H form invariant. Let $f(Z)$ be some analytic function and define a point transformation by

$$x = \mathrm{Re}[f(Z)] \quad y = \mathrm{Im}[f(Z)] \quad Z = X + iY.$$ (31)

Under this point transformation equation 29 stays form invariant, and the potential changes as

$$V(x, y) \to U(X, Y) = |f'(X + iY)|^2 V(\mathrm{Re}\,f(X + iY), \mathrm{Im}\,f(X + iY)).$$ (32)

The key element here is that we can multiply the whole expression without changing the constraint.

As an example,[4] let us take the Hamiltonian

$$H = \tfrac{1}{2}(p_x^2 + p_y^2) + ax^2 + by^2, \tag{33}$$

which is well known to be integrable (at any energy). If we take $f(Z) = Z^2$, then we find that the Hamiltonian

$$H = \tfrac{1}{2}(p_x^2 + p_y^2) + 4[ax^6 + (4b - a)x^4y^2 + (4b - a)x^2y^4 + ay^6] \tag{34}$$

is also integrable for any a and b, but only at zero energy. For arbitrary energy this system is known to be integrable only for the parameter values $a = 4b$, or $4a = b$, or $a = b$.

At zero energy there are also some other transformations that mix in p-dependence; see, for example, reference 28.

In equation 28 the right-hand side had a factor H that was put to zero. It is not necessary that H be a factor. What we need is just that there is a factor that remains zero during the motion. As an example, take the Fokas–Lagerstrom potential and invariant[29]

$$H = (p_x^2 + p_y^2)/2 + g(xy)^{-2/3}$$
$$I_2 = p_xp_y(xp_y - yp_x) + 2g(xp_x - yp_y)g(xy)^{-2/3}. \tag{35}$$

Let us now define a third function

$$I_3 = xp_x^3/(yp_y^3), \tag{36}$$

then

$$[H, I_3] = -I_2I_3/(xyp_xp_y), \tag{37}$$

that is, I_3 is a third invariant, if the second invariant I_2 has the value zero.

GENERALIZATIONS FROM 2 TO N DIMENSIONS

So far we have discussed only two-dimensional Hamiltonian models, which indeed is the place to start. Much less is known about the integrability of higher dimensional Hamiltonians. There are two basic techniques that produce such models that are integrable for any degree N.

In the method based on Lax pairs,[15] one constructs a matrix L (whose entries are functions of the p's and q's) with certain properties, and then a commuting set of functions are obtained from $I_n = \text{Tr}(L^n)$. If L is $N \times N$ with p's on the diagonal, this produces invariants whose leading order grows like n. For some other models L is $2N \times 2N$ with p's and $-p$'s on the diagonal, and then $I_n \approx p^{2n}$.

There is another technique that produces invariants of which N-2 are like angular momenta, the remaining two being the Hamiltonian and one special invariant. This method,[30] which we will next describe, is based on the integrability of a two-dimensional Hamiltonian with some special additive terms in the potential.

Assume that the Hamiltonian

$$H = \tfrac{1}{2}(p_x^2 + p_y^2) + V(x, y) + \tfrac{1}{2}ay^{-2} \tag{38}$$

is integrable for any value of the parameter a. Let us group together the terms $\frac{1}{2}(p_y^2 + ay^{-2})$, and since a is a constant, it can be interpreted as angular momentum and y the corresponding radial variable, that is,

$$\tfrac{1}{2}(p_y^2 + ay^{-2}) \to \tfrac{1}{2}\left(\sum_{i=1}^{N} p_i^2\right). \tag{39}$$

Thus the system generalizes from 2 to $N + 1$ dimensions. The precise rules of transcription for H and I_2 are as follows:

$$a \to \sum_{i=1}^{N} p_i^2 - p_y^2 \tag{40a}$$

$$yp_y \to \sum_{i=1}^{N} p_i y_i \tag{40b}$$

$$y^2 \to \sum_{i=1}^{N} y_i^2. \tag{40c}$$

To avoid ambiguities the replacements should be done in the order given.

If a potential can have both $\frac{1}{2}ay^{-2}$ and $\frac{1}{2}bx^{-2}$ as additional terms while preserving integrability, then extensions to two directions are possible. Note that the invariant may have to be changed extensively to allow for such additional terms. For example, the potential $V = y^4 + 12x^2y^2 + 16x^4$ is integrable with a quadratic invariant $I = (xp_y - yp_x)p_x + \cdots$, and the additional term $a\tfrac{1}{2}y^{-2}$ is allowed with minor modifications in I. However, the additional term $\frac{1}{2}bx^{-2}$ requires that the order of the invariant be doubled to $I = (xp_y - yp_x)^2 p_x^2 + \cdots$. For more details see reference 30.

CONCLUSIONS

We have given here a brief review of what is known about the integrability of two-degree-of-freedom flat-space Hamiltonian models. Quite a few integrable cases are already known, but further general results are probably difficult to come by.

It seems to me that most progress can be expected in the areas of nonpolynomial invariants, and of integrability at special values of energy or other invariants. We have here assumed time-independent Hamiltonians in flat space, but systems with more relaxed assumptions are also being studied with interesting results.

I am convinced that theoreticians working on various aspects of integrability would now appreciate input from users as to what kind of Hamiltonians should be studied and what kind of integrability is most useful in practical applications, for example, in astronomy.

ACKNOWLEDGMENTS

I would like to thank T. deZeeuv for bringing reference 24 to my attention, and J. H. Hunter, Jr., for discussions about the Fokas–Lagerstrom model. I have also benefited from discussions with G. Contopoulos, B. Grammaticos, C. Jung, M. Kruskal, A. Ramani, H.-J. Scholz, and H. Yoshida.

REFERENCES

1. RAMANI, A., B. GRAMMATICOS, T. BOUNTIS & H. YOSHIDA. 1988. The Painlevé property and singularity analysis of integrable and non-integrable systems. Phys. Rep. In press.
2. ZIGLIN, S. L. 1984. Branching of solutions and the nonexistence of first integrals in Hamiltonian mechanics. Func. Anal. Appl. **16**: 181; **17**: 6.
3. VAN MOERBEKE, P. 1985. Algebraic geometrical methods in Hamiltonian mechanics. Phil. Trans. R. Soc. London **A315**: 379.
4. HIETARINTA, J. 1987. Direct methods for the search of the second invariant. Phys. Rep. **147**: 87.
5. DARBOUX, G. 1901. Sur un problème de mécanique. Arch. Neerl. (ii) **6**: 371.
6. WHITTAKER, E. T. 1944. A Treatise on the Analytical Dynamics of Particles and Rigid Bodies. Dover. New York.
7. WINTERNITZ, P., J. A. SMORODINSKY, M. UHLIR & I. FRIS. 1967. On symmetry groups in quantum mechanics. Yad. Fiz. **4**: 625. (Sov. J. Nucl. Phys. **4**: 444.)
8. DORIZZI, B., B. GRAMMATICOS & A. RAMANI. 1983. A new class of integrable systems. J. Math. Phys. **24**: 2282.
9. ANKIEWICZ, A. & C. PASK. 1983. The complete Whittaker theorem for two-dimensional integrable systems and its application. J. Phys. A: Math. Gen. **16**: 4203.
10. THOMPSON, G. 1984. Darboux's problem of quadratic integrals. J. Phys. A: Math. Gen. **17**: 985.
11. SEN, T. 1985. Integrable potentials with quadratic invariants. Phys. Lett. **A111**: 97.
12. SEN, T. 1986. Integrable systems and Lie symmetries in classical mechanics. Ph.D. Thesis. University Microfilms International. Ann Arbor, Mich.
13. HOLT, C. R. 1982. Construction of new integrable Hamiltonians in two degrees of freedom. J. Math. Phys. **23**: 1037.
14. GRAMMATICOS, B., B. DORIZZI & A. RAMANI. 1983. Hamiltonians with higher-order integrals and the "weak-Painlevé" concept. J. Math. Phys. **25**: 3470.
15. OLSHANETSKY, M. A., & A. M. PERELOMOV. 1981. Classical integrable finite dimensional systems related to Lie-algebras, Phys. Rep. **71**: 313.
16. HALL, L. S. 1983. A theory of exact and approximate configurational invariants. Physica **8D**: 90.
17. GRAMMATICOS, B., B. DORIZZI & R. PADJEN. 1982. Painlevé property and integrals of motion for the Hénon-Heiles system. Phys. Lett. **A89**: 111.
18. HIETARINTA, J. 1983. Integrable families of Hénon-Heiles type and a new duality. Phys. Rev. **A28**: 3670.
19. RAMANI, A., B. DORIZZI & B. GRAMMATICOS. 1982. Painlevé conjecture revisited. Phys. Rev. Lett. **49**: 1539.
20. GRAMMATICOS, B., B. DORIZZI & A. RAMANI. 1983. Integrability of Hamiltonians with third- and fourth-degree polynomial potentials. J. Math. Phys. **24**: 2289.
21. BOGOYAVLENSKY, O. I. 1976. On perturbations of the periodic Toda lattice. Commun. Math. Phys. **51**: 210.
22. YOSHIDA, H. 1984. Integrability of generalized Toda lattice systems and singularities in the complex t-plane. *In* Nonlinear Integrable Systems—Classical Theory and Quantum Theory. M. Jimbo and T. Miwa, Eds.: 273. World Scientific. Singapore.
23. DORIZZI, B., B. GRAMMATICOS, R. PADJEN & V. PAPAGEORGIOU. 1984. Integrals of motion for Toda systems with unequal masses. J. Math. Phys. **25**: 2200.
24. YOSHIDA, H., A. RAMANI, B. GRAMMATICOS & J. HIETARINTA. 1987. On the nonintegrability of some generalized Toda lattices. Physica **144A**: 310.
25. DORIZZI, B., B. GRAMMATICOS, A. RAMANI & P. WINTERNITZ. 1985. Integrable Hamiltonian systems with velocity dependent potentials. J. Math. Phys. **26**: 307.

26. VANDERVOORT, P. O. 1979. Isolating integrals of the motion for stellar orbits in a rotating galactic bar. Astrophys. J. **232**: 91.
27. HIETARINTA, J. 1984. New integrable Hamiltonians with transcendental invariants. Phys. Rev. Lett. **52**: 1057.
28. HIETARINTA, J. 1985. How to construct Fokker-Planck and electromagnetic Hamiltonians from ordinary integrable Hamiltonians. J. Math. Phys. **26**: 1970.
29. FOKAS, A. S. & P. A. LAGERSTROM. 1980. Quadratic and cubic invariants in classical mechanics. J. Math Anal. Appl. **74**: 325.
30. GRAMMATICOS, B., B. DORIZZI, A. RAMANI & J. HIETARINTA. 1985. Extending integrable Hamiltonians from 2 to N dimensions. Phys. Lett. **A109**: 81.

Painlevé Expansions for Integrable and Nonintegrable Ordinary Differential Equations[a]

M. TABOR

Department of Applied Physics
Columbia University
New York, New York 10027

INTRODUCTION

A valuable test of integrability is proving to be the so-called Painlevé test, which has its origins in the classic work of Kovaleskaya on the rigid-body problem (a semi-historical review is given in reference 1). Here the basic idea is that for integrable systems the only movable singularities (i.e., singularities whose positions are initial condition dependent) exhibited by the solutions in the complex (time) domain are ordinary poles. In practice this test is easily implemented by demonstrating that the dependent variables can be expanded in the neighborhood of some arbitrary singularity (say, at t_0) in a Laurent series, that is

$$x(t) = \frac{1}{(t - t_0)^\alpha} \sum_{j=0}^{\infty} a_j (t - t_0)^j. \tag{1.1}$$

Here α is some "leading order" exponent—which must be integer for a single-valued expansion—and the a_j are a set of expansion coefficients of which the requisite number must be arbitrary in order that equation 1.1 be a local representation of the general solution. All sorts of interesting questions concerning this idea can arise, such as what happens if α is rational ("weak Painlevé") or if there are not enough arbitrary coefficients in the expansion ("singular solutions"), and, of course, why should such a test work?

As it stands the expansion 1.1 is purely "local" and would appear to contain no further information. However, as we shall describe, recent results that use the techniques developed for partial differential equations enable us to determine, from expansions like 1.1, the solutions to the problem in hand by finding the associated "Lax pair" from which we can then deduce the integrals of motion and the so-called algebraic curve. This procedure gives us some deep geometrical insights into the properties of integrable systems and why the Painlevé test works.

In addition one can ask what happens to the singularities of nonintegrable systems. Here one finds that they become multivalued in all sorts of interesting ways. Past work[2] has shown that when the singularities have complex exponents they cluster recursively

---

[a]This work was supported in part by Department of Energy Grant DE-FG02-84ER 13190. The author is an Alfred P. Sloan research fellow.

in the complex domain, forming self-similar natural boundaries. Our most recent work[3,4] has been concerned with commonly occurring logarithmic branch points. (These arise in physically important systems such as the Lorenz and Duffing equations.) Locally the solutions can be represented as so-called psi-series of the form

$$x(t) = \frac{1}{(t-t_0)^n} \sum_{j=0}^{\infty} \sum_{k=1}^{\infty} a_{jk}(t-t_0)^j ((t-t_0)^m \log(t-t_0))^k, \qquad (1.2)$$

where n and m are certain integers determined directly from the differential equations. As it stands, such a series might be regarded as being uninformative, if not downright unpleasant! However, we have developed a novel resummation technique that enables us to represent equation 1.2 in terms of known analytical functions. This enables us to achieve a type of "local integrability," that is, an explicit analytical representation of the solution in the neighborhood of a singularity. By piecing together such local solutions we believe that it is possible to provide a piecewise analytical representation of even chaotic trajectories. Furthermore, by considering the "resummed" psi-series as "pseudo Laurent" series, it is possible to use them, for special system parameter values, to construct certain integrals of motion.

INTEGRATING INTEGRABLE SYSTEMS

As outlined in the introduction, the Painlevé test for integrability is implemented by determining whether the dependent variable(s), say, $q(x)$, can be expanded locally in the neighborhood of a movable singularity x_0 as a Laurent series, that is,

$$q(x) = \frac{1}{(x-x_0)^n} \sum_{j=0}^{\infty} a_j (x-x_0)^j, \qquad (2.1)$$

where n is an (integer) leading order and a_j the expansion coefficients. The general solution will have as many arbitrary pieces of data (x_0 and various a_j) as the order of the system. Depending on the nonlinearities of the ordinary differential equations (ODE) other singularities can coexist with those of the general solution and may also have expansions of the form 2.1, but with less than the full complement of arbitrary coefficients. As will be discussed, these "singular branches" may contain much valuable information.

We say that a solution (of a differential equation) has the "*Laurent property*" if it can be represented in the form 2.1. A system (of differential equations) is said to have the "*Painlevé property*" if all the solution branches have the Laurent property and is then deemed to be integrable. (The role of movable essential singularities is still not understood.) For a while it seemed that this was as far as one could go, that is, the series 2.1 contained no further information about the actual solutions to the problem. However, it was recognized that the Laurent-like expansions developed by Weiss, Tabor, and Carnevale (WTC)[5] for partial differential equations involving the so-called *singular manifold* function could be profitably applied to ordinary differential equations as well. Thus one works with expansions of the form

$$q(x) = \frac{1}{\phi^n(x)} \sum_{j=0}^{\infty} q_j(x) \phi^j(x), \qquad (2.2)$$

where $\phi(x) = 0$ defines the singular manifold with the special choice $\phi(x) = x - x_0$ recovering the standard Laurent expansion 2.1. An important constraint on ϕ is that its gradient should not vanish on the singular manifold, that is, $\phi_x(x) \neq 0$ on $\phi(x) = 0$.

We illustrate the approach with the simple example of the stationary solution to the KdV equation, that is

$$q_{xx} + 3q^2 = 0. \tag{2.3}$$

Substitution of the expansion 2.2, truncated at $O(\phi^0)$, that is, up to the coefficient q_2, yields

$$q_{xx} + 3q^2 = \frac{1}{\phi^2}\{-8\phi_x\phi_{xxx} - 12\phi_x^2 q_2 + 6\phi_{xx}^2\}$$

$$+ \frac{1}{\phi}\{2\phi_{xxxx} + 12\phi_{xx}q_2\} + q_{2xx} + 3q_2^2, \tag{2.4}$$

which on making the "squared eigenfunction" substitution (discussed in reference 6)

$$\phi_x = \Psi^2, \tag{2.5}$$

becomes

$$q_{xx} + 3q^2 = -\frac{4\Psi^2}{\phi^2}\left\{\int^x \Psi(4\Psi_{xxx} + 6q_2\Psi_x + 3q_{2x}\Psi)\,dx'\right\}$$

$$+ \frac{4\Psi}{\phi}\{4\Psi_{xxx} + 6q_2\Psi_x + 3q_{2x}\Psi\}$$

$$+ q_{2xx} + 3q_2^2. \tag{2.6}$$

The usual WTC prescription is to set each order of ϕ equal to zero, which would here give the system of equations

$$\Psi_{xx} + q_2\Psi = \lambda\Psi \tag{2.7a}$$

$$4\Psi_{xxx} + 6q_2\Psi_x + 3q_{2x}\Psi = 0 \tag{2.7b}$$

$$q_{2xx} + 3q_2^2 = 0. \tag{2.7c}$$

Although this system of equations is self-consistent, it is not general enough to integrate equation 2.7c, and can only provide special solutions. A crucial new observation, made in recent work with Newell and Zeng,[6] is that the original WTC prescription is too restrictive and instead one should think of each order of ϕ being zero *modulo some function of ϕ*. Thus in this case if we

(1) add an amount $4y\phi_x\phi$ at $O(\phi^{-2})$, and
(2) subtract the identical amount $4y\phi_x$ at $O(\phi^{-1})$, where y is an additional free parameter, we obtain instead of equation 2.7b

$$4\Psi_{xxx} + 6q_2\Psi_x + 3q_{2x}\Psi = y\Psi, \tag{2.8}$$

which together with equation 2.7a is the correct Lax pair for equation 2.7c.

A less trivial example is an integrable case of the Henon–Heiles system, namely

$$q_{xx} = -3q^2 - \tfrac{1}{2}p^2 \qquad (2.9a)$$

$$p_{xx} = -qp. \qquad (2.9b)$$

WTC expansions of q and p, for the general solution, truncated at $O(\phi^0)$, take the form

$$q = 2\frac{\partial^2}{\partial x^2}\log\phi + q_2, \qquad p = \frac{p_0}{\phi} + p_1, \qquad (2.10)$$

and when these are substituted in the equations 2.9, one is able to obtain the Lax pair[6]

$$\Psi_{xx} + (q_2 + \lambda)\Psi = 0 \qquad (2.11a)$$

$$y\Psi + 4\Psi_{xxx} + 6q_2\Psi_x + 3q_{2x}\Psi - \tfrac{1}{4}p_1 \int^x p_1\Psi\, dx' = 0, \qquad (2.11b)$$

where again we use the relation $\phi_x = \Psi^2$. We believe that this result is the first example of the WTC method yielding a completely new Lax pair (apparently associated with the loop algebra of $A_3^{(2)}$).

It is convenient to rewrite equations 2.11, in system form, as

$$W_x = PW \qquad y\lambda W = QW \qquad (2.12)$$

with

$$W = \begin{pmatrix} \Psi \\ \Psi_x \end{pmatrix} \qquad P = \begin{pmatrix} 0 & 1 \\ -\lambda - u & 0 \end{pmatrix}$$

$$Q = \begin{pmatrix} \lambda u_x + \tfrac{1}{4}vv_x & 4\lambda^2 - 2\lambda u - \tfrac{1}{4}v^2 \\ -4\lambda^3 - 2\lambda^2 u - \lambda(u^2 + \tfrac{1}{4}v^2) + \tfrac{1}{4}v_x^2 & -\lambda u_x - \tfrac{1}{4}vv_x \end{pmatrix} y, \qquad (2.13)$$

where we have set $u = q_2$ and $v = p_1$. The solvability condition of equation 2.12 is

$$Q_x = [P, Q], \qquad (2.14)$$

which, when written in component form, gives equations 2.9 (with q and p replaced by u and v, respectively). Further, if W is a fundamental solution matrix of equation 2.12, then the solution of equation 2.14 can be conveniently written as

$$Q = WQ_0W^{-1}, \qquad (2.15)$$

with Q_0 independent of x. As a consequence of equation 2.15, the characteristic polynomial of Q

$$\det(Q - y\lambda I) = 0, \qquad (2.16)$$

which is the condition that equation 2.12 has a nontrivial solution, is independent of x.

In terms of y, λ, v, and u, equation 2.16 is the algebraic curve,

$$\lambda^2 y^2 = -16\lambda^5 + 2H\lambda^2 + G\lambda, \qquad (2.17)$$

where

$$H = \tfrac{1}{2}(U^2 + V^2) + \tfrac{1}{2}uv^2 + u^3 \qquad (2.18)$$

is the Hamiltonian of the system 2.9 written in the canonically conjugate coordinate pairs u, $U = u_x$ and v, $V = v_x$ and

$$G = \tfrac{1}{4}v^2 (u^2 + \tfrac{1}{4}v^2) + \tfrac{1}{2}V(vU - uV). \qquad (2.19)$$

Because equation 2.17 is independent of time, we can identify G as the second constant of the motion for the flow 2.9 generated by the Hamiltonian H. Terms G and H are in involution under the canonical Poisson bracket

$$\{G, H\} = \frac{\partial G}{\partial V}\frac{\partial H}{\partial v} + \frac{\partial G}{\partial U}\frac{\partial H}{\partial u} - \frac{\partial G}{\partial v}\frac{\partial H}{\partial V} - \frac{\partial G}{\partial u}\frac{\partial H}{\partial U}. \qquad (2.20)$$

The choice of auxiliary variables, which leads to the identification of the angle variables corresponding to the actions H and G, is most conveniently made using the matrix from the Lax equation 2.12. The new variables are μ_1, μ_2 given by

$$\mu_1 + \mu_2 = \tfrac{1}{2}u$$

$$\mu_1\mu_2 = -\tfrac{1}{16}v^2, \qquad (2.21)$$

which are the zeros of the (1, 2) element, $4\lambda^2 - 2\lambda u - \tfrac{1}{4}v^2$, of Q (see reference 6). The reader can also verify that this choice of variables separates the associated Hamilton–Jacobi equation. (I recommend strongly that the reader consult the works of Ercolani and Siggia[7]—see below—on these ideas.) Standard techniques of algebraic geometry can then yield an explicit linearization of the equations of motion on the associated Jacobi variety. A careful analysis of the rational solutions of equations 2.9 shows that the singular solution (lower branch) exhibited by these equations is simply a reexpansion of the general solution (principal branch) about a location on the singular manifold $\phi = 0$ at which $\phi_x = 0$. (A detailed discussion of this is given in reference 6.)

Despite these successes important issues remain to be resolved. One vexed issue concerns the properties of systems with rational branch points. In the case of the Henon–Heiles system, one of the integrable cases was identified[8] by relaxing the Painlevé test to include such singularities. However, in that case, the branch point could be transformed away by a simple change of dependent variable. These results led Ramani et al.[9] to suggest that a "weak Painlevé" property, that is, nontransformable branch points, could still identify integrable systems, and a number of integrable Hamiltonians were constructed with this behavior. However, subsequent work[10] showed that some of these cases were separable, thereby making the branch points transformable. Another important issue is to determine, at least for algebraically integrable Hamiltonians (i.e., those with integrals that are algebraic functions of the canonical variables), the order of the integrals of motion (if they exist). Interestingly enough, this information does not require a complete solution to the problem, but may be contained in the singular branches of the solutions. Indeed, this ties in with the other major issue, namely, the information contained in these branches. It would appear, at least for certain classes of Hamiltonian, that systems of N degrees of freedom exhibit N solution branches. One of these, as we have described, corresponds to the general solution and the associated Laurent series exhibits $2N - 1$ arbitrary constants. The other branches have less free constants, with the most singular branch exhibiting N arbitrary parameters. Recent work of Ercolani and Siggia,[7] based on the analytic properties of the Hamilton–Jacobi equation (as well as our own on the stationary KdV hierarchy[6]) indicates that much information on the integrals is contained in this most singular branch. For systems with polynomial integrals it would appear that the order of the polynomial is easily determined by the so-called "resonance," or "Kovalevskaya exponent," structure, namely, the powers of the independent variable at which the free constants enter the Laurent expansions. This result could be rather useful, since it suggests an alternative route (to our determination of the algebraic curve) for finding the integrals. Furthermore, this may help with resolving the "weak Painlevé" issue. That is, if the structure of the most singular branch indicates the possibility of polynomial integrals (so far most weak Painlevé systems are of this type), we can probably allow the solution(s) to exhibit the weak property. (At this stage, however, we know little about the analytic structure of systems with transcendental integrals.)

"INTEGRATING" NONINTEGRABLE SYSTEMS

Systems with multivalued movable singularities fail the Painlevé test (as it currently stands) and are presumably nonintegrable. As indicated in the Introduction the local solution must now be represented in the form of a psi-series. The main types of multivaluedness are logarithmic, irrational, or complex, with the case of rational branch points ("weak Painlevé") having been discussed in the previous section.

Naturally one would like to know what information can be extracted from these psi-series. An example is provided by the work of Chang et al.[2] on certain nonintegrable regimes of the Henon–Heiles system. Here the singularities are complex and the psi-series are of the form

$$q(x) = \frac{1}{x^2} \sum_j \sum_k a_{jk} x^j (x^\beta)^k + \text{c.c.}, \tag{3.1}$$

where β is a complex number. These workers observed that a substitution of the form

$$q(x) = \frac{1}{x^2} \Theta(z), \tag{3.2}$$

where $\Theta(z)$ is some function of the variable $z = x^\beta$, into the original equations of motion gives, in the limit $x \to 0$, a differential equation for Θ that has the same analytic structure as the original equations. This is, in effect, a "renormalization" of the equations of motion in the neighborhood of a given singularity (here assumed to be at $x_0 = 0$). By mapping back the singularities from the z plane to the x plane, as determined by the transformation

$$z = x^\beta \tag{3.3}$$

it is fairly easy to demonstrate (in confirmation with the numerical results) that the singularities in the x domain cluster in self-similar spirals.

Although the psi-series for logarithmic singularities had been studied (in the context of the Lorenz equations) by Tabor and Weiss[11] sometime ago, it was not understood how the singularities might cluster for this type of multivaluedness. Recent work with Fournier and Levine[3] has demonstrated that for psi-series of the form

$$q(x) = \frac{1}{x^n} \sum_j \sum_k a_{jk} x^j (x^m \ln x)^k \tag{3.4}$$

the singularities in the complex x domain cluster recursively in the form of m-armed stars. As will be discussed below we were able to push the earlier analysis of reference 11 much further and obtain some certain asymptotic expansions that provide an effective "local integration" of the equations of motion.

A remarkable feature of "renormalizing" substitutions of the form 3.2 is that in some cases, such as certain nonintegrable regimes of the Henon–Heiles system, they give back a rescaled version of the same equation, whereas in other cases they yield *different* equations. For example, in the case of the Duffing oscillator (here ' denotes d/dx)

$$y'' + \lambda y'' + \tfrac{1}{2} y^3 = \epsilon g(x) \tag{3.5}$$

the substitution

$$y(x) = \frac{1}{x} \Theta(z) \tag{3.6}$$

where

$$z = x^4 \ln x \tag{3.7}$$

gives an equation for Θ that can be integrated in terms of elliptic functions—in fact, the resulting equation is the integrable part of equation 3.5, namely, of the form $f'' + \tfrac{1}{2} f^3 = 0$. In a way, this is rather extraordinary, since it means that in the neighborhood of a given singularity x_0 (here we have set $x_0 = 0$) we have an explicit analytical representation of the solution to equation 3.5, which is traditionally regarded as nonintegrable. A similar result was obtained for the Lorenz equations[11] where, depending on whether the psi-series involves terms of the form $x^2 \ln x$ or $x^4 \ln x$ (this depends on the system parameters), the corresponding equation for Θ could be integrated in terms of different types of elliptic functions. These observations suggest that nonintegrable systems can be classified according to whether or not they have this "local integrability" property. This seems to be rather a deep concept, but we have, as yet, little understanding of its significance.

Additional, important results have recently been obtained for the Duffing equation 3.5.[3] Here we have found that the substitution of equation 3.6 is in fact, just the first term in an asymptotic expansion of the form

$$y(x) = \sum_{j=0}^{\infty} \Theta_j(z) x^{j-1}, \tag{3.8}$$

where the whole set of $\Theta_j(z)$ can be determined analytically. The leading term Θ_0 captures the essential nonlinearities of the system, and the subsequent $\Theta_j (j \geq 1)$ satisfy linear equations that can all be integrated in terms of Lamé functions. If the preceding expansion is compared with the original psi-series

$$y(x) = \sum_{j=0}^{\infty} \sum_{k=1}^{\infty} a_{jk} x^{j-1} (x^4 \ln x)^k, \qquad (3.9)$$

we realize that each Θ_j is just

$$\Theta_j(z) = \sum_k a_{jk} z^k \qquad (3.10)$$

($z = x^4 \ln x$), which are the generating functions for the coefficient sets a_{jk} ($k = 1, \ldots, \infty$). Thus the series 3.8 constitutes a systematic resummation of the psi-series 3.9. It would appear that this new type of series have a very rich content whose properties have yet to be fully investigated.

These new insights into the nature of logarithmic singularities have enabled us to probe a number of important questions. For example, in the case of the Lorenz equations

$$X' = \sigma(Y - X) \qquad (3.11a)$$

$$Y' = -XZ + RX - Y \qquad (3.11b)$$

$$Z' = XY - BZ, \qquad (3.11c)$$

there are certain values of the system parameters σ, B, R for which the equations, although nonintegrable, possess one integral of the motion. These integrals have been identified by Kus.[12] We have been able to demonstrate that the existence of these integrals coincides with a subtle change in singularity clustering and, furthermore, the associated "resummed" psi-series can actually be used to construct these integrals.[4] Very briefly, the variables X, Y, Z are typically represented as the resummed psi-series

$$X = \sum_{j=0}^{\infty} \Theta_j(z) x^{j-1} \qquad (3.12a)$$

$$Y = \sum_{j=0}^{\infty} \Phi_j(z) x^{j-2} \qquad (3.12b)$$

$$Z = \sum_{j=0}^{\infty} \Psi_j(z) x^{j-2}, \qquad (3.12c)$$

where Θ_j, Φ_j, Ψ_j are the generating functions previously described, and the variable z is either $x^2 \ln x$ or $x^4 \ln x$. For the case of $z = x^2 \ln x$, the Θ_0, Φ_0, Ψ_0 (and hence, in fact, the generating functions for all j) can be solved in terms of Jacobi elliptic functions; whereas, for $z = x^4 \ln x$ the solution is in terms of the Lemniscate function. In the former case, the shape of the singularity lattice of the elliptic function is determined by certain combinations of the system parameters. For special parameter sets the elliptic function collapses to either a sin or a sinh. The overall singularity clustering is then much simplified and the functions Θ_j, Φ_j, Ψ_j take on particularly simple forms. In these cases, it is possible to make polynomial combinations of the X, Y, and Z such that the singular parts of the "pseudo-Laurent" series 3.12 cancel. The resulting functions of X,

Y, and Z are thus *entire functions* of x, and hence, by Liouville's theorem, are constant. In short, they turn out to be precisely the integrals identified by Kus!

REFERENCES

1. TABOR, M. 1984. Nature **310**: 277.
2. CHANG, Y. F., J. M. GREENE, M. TABOR & J. WEISS. 1983. Physica **8D**: 183.
3. FOURNIER, J. D., G. LEVINE & M. TABOR. 1988. Singularity clustering in the Duffing oscillator. J. Phys. A **21**: 33.
4. LEVINE, G. & M. TABOR. 1988. Integrating the nonintegrable: Analytic structure of the Lorenz equations revisited. Accepted for publication in Physica D.
5. WEISS, J., M. TABOR & G. CARNEVALE. 1983. J. Math. Phys. **24**: 522.
6. NEWELL, A. C., M. TABOR & Y. ZENG. 1987. A unified approach to Painlevé expansions. Physica **29D**: 1.
7. ERCOLANI, N. & E. SIGGIA. 1986. The Painlevé property and integrability. Phys. Lett. **A119**: 112.
8. CHANG, Y. F., M. TABOR & J. WEISS. 1982. J. Math. Phys. **23**: 531.
9. RAMANI, A., B. DORIZZI & B. GRAMMATICOS. 1982. Phys. Rev. Lett. **49**: 1539.
10. ANKIEWICZ, A. & C. PASK. 1983. J. Phys. A **16**: 4203.
11. TABOR, M. & J. WEISS. 1981. Phys. Rev. A **24**: 2157.
12. KUS, M. 1983. J. Phys. A. **16**: L689.

Solving the Vlasov Equation in General Relativity by Particle Simulation[a]

STUART L. SHAPIRO AND SAUL A. TEUKOLSKY

Center for Radiophysics and Space Research
Departments of Astronomy and Physics
Cornell University
Ithaca, New York 14853

INTRODUCTION

General relativity provides an interesting and nontrivial set of equations for the exploration of nonlinear dynamics. Only now, with the advent of supercomputers, can we begin to explore the character of the full range of solutions.

We have recently shown that it is possible to integrate the full Einstein equations for the dynamical motion of an arbitrary spherical, *collisionless* configuration in general relativity (references 1, 2, 3, 4, hereafter Papers I, II, III and IV, respectively). We can track the dynamical evolution of a relativistic star cluster on the computer by combining the techniques of numerical relativity with those of N-body particle simulations. Our resulting method is even able to handle epochs characterized by total gravitational collapse leading to the formation of a black hole. Our general relativistic Vlasov integrations follow the formation and growth of the black hole accurately without the appearance of numerical or physical singularities.

In Paper I we presented some historical background to this problem, together with a detailed discussion of the computational method and a battery of test-bed calculations that help assess the reliability of our code. In Paper II we applied the code to address several long-standing theoretical issues in relativistic stellar dynamics, including the stability of relativistic star clusters in dynamical equilibrium, the collapse of unstable star clusters to black holes, and relativistic "violent relaxation." Based on our findings, we proposed in Paper III a plausible scenario for the origin of quasars and active galactic nuclei (AGNs) via the collapse of dense star clusters embedded in galactic nuclei to supermassive black holes. In Paper IV we explored alternative coordinate choices, ideally suited for evolving the very centrally condensed clusters most likely to form in Nature.

These computations are an example of a problem that requires extensive use of graphical display to visualize the dynamical behavior predicted by the simulation. We have recently produced a computer-generated color movie that dramatically illustrates the collapse of unstable clusters to black holes.

We thus feel that "relativistic stellar dynamics on the computer" is an endeavor

---

[a]This work was supported in part by National Science Foundation Grants AST 84-15162 and PHY 86-03284 at Cornell University. Computations supporting the research were performed on the Cornell Production Supercomputer Facility, which is supported in part by the National Science Foundation and IBM corporation.

ripe for computational and astrophysical exploration. Probing this area provides fresh insights into nonlinear dynamics and general relativity. It establishes a new laboratory in which to test numerical algorithms for the construction of space–times. It also opens up some promising avenues for the formation of supermassive black holes in Nature, which are believed to be the likely sources of energy in quasars and AGNs.

This paper is a summary of the highlights of Papers I–IV.

MOTIVATION

The motivation for solving the collisionless gas problem in general relativity is at least threefold: computational, theoretical, and astrophysical. Computationally, this problem falls into two broad categories of research now being hotly pursued by computational physicists: *nonlinear dynamics* and *field theory on a space–time lattice*. Interest in these areas is mushrooming, in part because of the availability of new computer hardware and in part because of the development of new numerical algorithms. It is also mushrooming because progress achieved in solving any one problem in any particular area often leads to progress in solving other problems in ostensibly different areas. This phenomenon occurs because, when expressed in computational terms, many different problems share a common numerical structure.

Relativistic stellar dynamics on the computer poses a considerable computational challenge. The equations are coupled, nonlinear, many-body, multidimensional, time-dependent differential equations. These are just the attributes that excite a computational physicist!

Working with *collisionless* matter has several advantages over *fluid* systems for doing numerical relativity. The collisonless matter equations are *ordinary* differential equations (the geodesic equations), while hydrodynamical equations are *partial* differential equations. Furthermore, collisionless matter is not subject to shocks or other hydrodynamical discontinuities. These pathologies require special care in numerical simulations. Their absence allows one to focus all of the computational effort on solving Einstein's field equations. Consequently, the collisionless matter environment provides a golden opportunity to experiment with different numerical algorithms, coordinate choices, and so on. It is the Camelot for searching for the Holy Grail of numerical relativity—a code that simultaneously

- avoids singularities
- handles black holes
- maintains high accuracy
- runs forever.

Turning to the theoretical motivation, this work addresses several long-standing issues in relativistic stellar dynamics:

- stability of relativistic star clusters in dynamic equilibrium
- binding energy criterion for stability
- nonlinear evolution of unstable configurations and the collapse of star clusters to black holes
- relativistic violent relaxation.

Previous treatments of the stability of relativistic star clusters[5-7] were restricted to linear perturbation theory. No simple binding energy criterion for the onset of instability has yet been rigorously derived, in contrast to the case of fluid stars. (Ipser[8] has presented a sufficient, but not necessary, criterion). The final fate of unstable clusters and their collapse to black holes has been discussed only in qualitative terms up till now.[9]

As for the astrophysical motivation, the most important application of our work is finding plausible scenarios for the formation of supermassive black holes in Nature. Such objects are believed to be the engines that power quasars and AGNs. Understanding how supermassive black holes form may thus explain the origin of these phenomena.

Supermassive black holes are also convenient sites for hiding "dark matter" in the universe. Knowing how they arise might shed light on the problem of the missing mass.

If these are not reasons enough for studying relativistic stellar dynamics, we can always appeal to Authority:

> ... the collapse of a system of particles is very complicated. The problem is more difficult than the nonstationary hydrodynamical problem.... It is very desirable to perform calculations for such a model.
> —*Zel'dovich and Podurets*[9]

> A numerical investigation of such a collapse is badly needed, but no one up to now has had the fortitude to attempt it.
> —*Zel'dovich and Novikov*[10]

PHYSICAL PICTURE

Consider a sphere drawn in the interior of the collisionless matter distribution (FIG. 1). We imagine that the sphere is densely and uniformly covered with an infinite number of particles, each with infinitesimal rest mass. At any point on the sphere particles move in both the radial and transverse directions. To preserve spherical symmetry, we require that their motion be isotropic in the transverse plane. Thus, while individual particles may have large angular momentum about the cluster center, the total angular momentum summed over all the particles is zero.

The restriction to spherical symmetry reduces the number of phase-space degrees of freedom we have to keep track of. In *coordinate* space the only nontrivial dynamical variable is the radius r of a particle. In *velocity* space the nontrivial dynamical variables are the radial and transverse velocities, u^r and $u^\perp$.

NEWTONIAN LIMIT

Before considering the fully relativistic problem, it is instructive to examine how the problem is solved in Newtonian physics. For the motion of nonrelativistic particles, the metric is just

$$ds^2 = -(1 + 2\Phi)dt^2 + dr^2 + r^2 d\Omega^2, \tag{4.1}$$

where Φ is the Newtonian potential. The matter moves according to the geodesic equations, which are simply Newton's laws of motion:

$$\frac{d\mathbf{x}}{dt} = \mathbf{u},$$

$$\frac{d\mathbf{u}}{dt} = -\nabla\Phi, \qquad (4.2)$$

where $\mathbf{x}$ and $\mathbf{u}$ are the position and velocity 3-vectors of each particle. Because of spherical symmetry, these equations simplify to

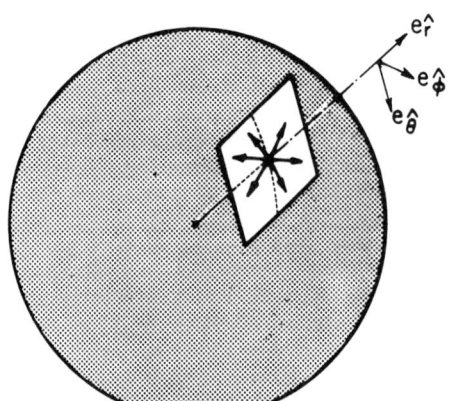

FIGURE 1. Schematic representation of the distribution of particles in a spherical star cluster.

$$\frac{dr}{dt} = u_r,$$

$$\frac{du_r}{dt} = -\Phi_{,r} + \frac{u_\phi^2}{r^3},$$

$$u_\theta = 0 \quad \text{(orbit confined to a plane, e.g., } \theta = \pi/2\text{)},$$

$$\frac{du_\phi}{dt} = 0 \quad \text{(conservation of angular momentum).} \qquad (4.3)$$

These equations are integrated for every particle for a small timestep. The new particle positions then yield the rest-mass density ρ at the new time

$$\rho = \sum_{\text{all particles}} mn, \qquad (4.4)$$

where m is the particle rest mass and n the number density. The rest-mass density serves as the *source* term for the gravitational field equation, which in this case is simply Poisson's equation

$$\nabla^2 \Phi = 4\pi G \rho. \tag{4.5}$$

Solving this equation gives the self-consistent gravitational field at the new time. The new potential is then inserted in the particle equations of motion and the process is repeated. This approach to evolving a self-gravitating, N-body system is known as a *mean-field, particle simulation scheme*.

RELATIVISTIC EQUATIONS

Solving the relativistic problem is similar in spirit to the Newtonian approach just described. The Newtonian particle equations of motion are replaced by the geodesic equations of general relativity, while Poisson's equation is replaced by Einstein's equations for the gravitational field.

Einstein's equations for the gravitational field are by no means trivial. The textbook form of these equations,

$$G_{\mu\nu} = \frac{8\pi G}{c^4} T_{\mu\nu}, \tag{5.1}$$

where $G_{\mu\nu}$ is the Einstein tensor and $T_{\mu\nu}$ is the stress-energy tensor, intertwines space and time. This form is not suitable for performing a computer evolution. We need to recast the equations in the form of an *initial-value problem:* given the state of the system over all space at one instant of time t, we require equations that determine the state of the system over all space at the next instant $t + dt$. The new state can then be used as initial data to continue the integration to the next instant, and so on.

The required splitting of space and time in general relativity is given by the ADM or $3 + 1$ decomposition of Einstein's equations.[11] In this decomposition the original form, equation 5.1, of Einstein's equations is split into two kinds of equations: *constraint* equations and *evolution* equations. The constraint equations contain no time derivatives and relate field variables at a given instant of time. The evolution equations are first-order differential equations in time that propagate the initial data to the next instant of time.

This kind of decomposition is by no means unique to general relativity. Indeed, Maxwell's equations of electromagnetism are already in this form: the constraint equations of electromagnetism are

$$\nabla \cdot \mathbf{E} = 4\pi \rho_e \qquad \nabla \cdot \mathbf{B} = 0, \tag{5.2}$$

while the evolution equations are

$$\frac{\partial \mathbf{E}}{\partial t} = c\nabla \times \mathbf{B} - 4\pi \mathbf{J},$$

$$\frac{\partial \mathbf{B}}{\partial t} = -c\nabla \times \mathbf{E}. \tag{5.3}$$

Here $\mathbf{E}$ is the electric field, $\mathbf{B}$ is the magnetic field, ρ_e is the electric charge density, $\mathbf{J}$ is the current density, and c is the speed of light. In a numerical simulation of Maxwell's equations, one would start with **E** and **B** fields that satisfy the constraint equations 5.2. The subsequent evolution of these fields is governed by equations 5.3. The whole procedure is self-consistent: the constraint equations are always guaranteed to hold if they are satisfied initially and if the fields are evolved according to equations 5.3.

Note that the Newtonian limit of the ADM decomposition yields one constraint equation, which is just Poisson's equation, equation 4.5. There are no evolution equations—Newtonian gravitation is not a truly dynamical field.

In full general relativity, the ADM equations are similar in spirit to Maxwell's equations, but more complicated. There are more fields and equations, since general relativity describes a tensor field, while Maxwell's equations describe vector fields. More importantly, the field equations in general relativity are *nonlinear*. This makes the equations much more difficult to solve, and yields solutions exhibiting new types of physical behavior.

In solving the ADM equations, we adopt the isotropic form of the metric:

$$ds^2 = -(\alpha^2 - A^2\beta^2)dt^2 + 2A^2\beta dr\, dt + A^2(dr^2 + r^2 d\theta^2 + r^2 \sin^2\theta d\phi^2). \quad (5.4)$$

Here α and β are the lapse and shift functions of ADM. We use units with $c = G = 1$.

Matter Equations

The matter is described by the stress-energy tensor for a collisionless swarm of particles,

$$T^{\mu\nu} = \sum_A m_A n_A u_A^\mu u_A^\nu. \quad (5.5)$$

Here m_A is the rest mass of particles of type A, u_A^μ is their 4-velocity, and n_A is their comoving number density.

The equation of motion for each particle is the geodesic equation

$$u^\mu_{;\nu} u^\nu = 0. \quad (5.6)$$

Once again these equations simplify in spherical symmetry, yielding

$$\frac{dr}{dt} = \frac{\alpha u_r}{A^2(\alpha u^0)} - \beta,$$

$$\frac{du_r}{dt} = -(\alpha u^0)\alpha_{,r} + u_r \beta_{,r} + \frac{\alpha u_r^2}{(\alpha u^0)} \frac{A_{,r}}{A^3} + \frac{\alpha u_\phi^2}{(\alpha u^0)}\left(\frac{1}{r^3 A^2} + \frac{A_{,r}}{r^2 A^3}\right),$$

$$u_\theta = 0,$$

$$\frac{du_\phi}{dt} = 0. \quad (5.7)$$

The normalization of the 4-velocity,

$$u_\mu u^\mu = -1, \quad (5.8)$$

gives

$$\alpha u^0 = \left(1 + \frac{u_r^2}{A^2} + \frac{u_\phi^2}{r^2 A^2}\right)^{1/2}. \tag{5.9}$$

Thus, given the metric at any time t, equations 5.7 and 5.9 are integrated for the new positions and 4-velocities of the particles at $t + \Delta t$.

In the Newtonian limit, $A \to 1$, $\alpha \to 1$, $\alpha_{,r} \to \Phi_{,r}$, $\beta \to 0$, and $u_0 \to 1$. We thus recover the Newtonian equations of motion discussed in the previous section.

Source Terms

Given the particle positions and velocities, one can calculate the source terms needed in the field equations:

$$\rho = \sum_A m_A n_A (\alpha u_A^0)^2,$$

$$t_r = -\sum_A m_A n_A (\alpha u_A^0) u_{r(A)},$$

$$T = -\sum_A m_A n_A,$$

$$S_{rr} = \sum_A m_A n_A u_{r(A)} u_{r(A)}. \tag{5.10}$$

Field Equations

One of the challenges of numerical relativity is to find a good choice of coordinates so that one can actually solve for the complete evolution of a system without encountering singularities. We have explored two choices of time coordinate. The first is the time coordinate defined by the *maximal slicing* condition. Mathematically, maximal slicing is defined by setting the trace of the extrinsic curvature of the $t = $ constant slices to zero:

$$\text{Tr } K_{ij} \equiv K_i^i = 0. \tag{5.11}$$

Taking the trace of the evolution equation for $\partial_t K_{ij}$ then leads to an elliptic equation for the lapse function α, equation 5.18 below.

A different time-slicing condition has been proposed by Eardley[12] and by Bardeen and Piran.[13] This is the *polar slicing* condition: set the trace of the *transverse* part of K_{ij} to zero, that is, the part orthogonal to the radial direction. This can be written

$$K_T \equiv K_\theta^\theta + K_\phi^\phi = 0. \tag{5.12}$$

The primary motivation for polar slicing was to find a coordinate condition that would make it easier to compute the amount of gravitational radiation emitted during nonspherical gravitational collapse. It was also hoped that this condition might have better singularity avoidance than maximal slicing when a black hole forms. This hope arises because polar slicing avoids regions of space–time containing trapped surfaces in spherical symmetry, and trapped surfaces signal the presence of an event horizon and the impending formation of a singularity. We use our code to explore whether these expectations are actually met.

It is convenient to use the quantity K_T to write equations valid in both maximal and polar gauges:

$$K_T = \begin{cases} 0, & \text{polar slicing,} \\ -K_r^r, & \text{maximal slicing.} \end{cases} \quad (5.13)$$

Then the evolution equation for the metric coefficient A is

$$A_{,t} = \beta\left(A_{,r} + \frac{A}{r}\right) - \frac{1}{2}\alpha A K_T. \quad (5.14)$$

The Hamiltonian constraint becomes

$$\frac{1}{r^2}\frac{\partial}{\partial r}\left(r^2 \frac{\partial}{\partial r} A^{1/2}\right) = -\frac{A^{5/2}}{4}\left(8\pi\rho + \frac{3K_T^2}{4}\right). \quad (5.15)$$

The shift β is given by the equation

$$\beta = -r \int_r^\infty \alpha\left(K_r^r - \frac{1}{2}K_T\right)\frac{dr}{r}. \quad (5.16)$$

The momentum constraint determines the radial component of the extrinsic curvature:

$$K_r^r = \begin{cases} -\dfrac{8\pi}{A^3 r^3}\int_0^r A^3 r^3 t_r\, dr, & \text{maximal,} \\[2mm] -\dfrac{4\pi r t_r}{1 + rA_{,r}/A}, & \text{polar.} \end{cases} \quad (5.17)$$

The lapse equation for maximal slicing is

$$\frac{\partial}{\partial r}\left(Ar^2 \frac{\partial}{\partial r}\alpha\right) = \alpha A^3 r^2 \left[\frac{3}{2}(K_r^r)^2 + 8\pi\rho + 4\pi T\right], \quad (5.18)$$

while for polar slicing we get

$$\alpha = \frac{A_{max}\alpha_{max}}{A}\exp\left[\frac{1}{2}\int_{r_{max}}^r \frac{r(A_{,r}/A)^2 + 8\pi r S_{rr}}{1 + rA_{,r}/A}\, dr\right]. \quad (5.19)$$

Boundary Conditions

The boundary conditions on the metric variables are discussed in Paper I for maximal slicing and Paper IV for polar slicing. In brief, we match to asymptotic flatness at large r and impose regularity at the origin.

Diagnostics

A number of self-consistency checks and diagnostic parameters are described in Papers I and IV. These include: conservation of the total mass-energy M; conservation of particle energy for stable, equilibrium clusters; null geodesic equations for light rays, used to map out the position of the event horizon; the criterion for the presence of trapped surfaces; and probes of the velocity distribution.

α-FREEZING

Numerically, the preceding equations yield *very* accurate space–times for the most part. However, as discussed in Paper IV, some of our integrations of collapsing clusters terminate well before the exterior space–time surrounding a growing, central black hole reaches a final stationary state. Thus it is not always possible for us to determine exactly what fraction of the total mass of a cluster ultimately forms a black hole and what fraction remains outside in orbit about the central hole.

The situation is quite different for *fluids* where, once a black hole forms at the center of a collapsing configuration, the entire mass ultimately flows inside the event horizon. In contrast to a collisionless gas, a fluid cannot form a nonsingular, *static* configuration consisting of a central black hole embedded in an ambient gas cloud—pressure forces would be infinite at the horizon.

Not surprisingly, the problem of determining the final black hole mass is most severe when we evolve clusters with appreciable *central concentration*. Such configurations by construction are characterized by enormous dynamic range, with the dynamical (orbit) timescale in the central core significantly shorter than the dynamical timescale in the outer halo. Accurately following the orbits of the central stars near the event horizon on timescales sufficiently long to track the outer halo stars is crucial for determining the final fate of the configuration. This requirement imposes the ultimate limitation on the ability of our code to integrate arbitrarily far into the future.

Overcoming the preceding limitation is not merely of pedagogic interest. For as we demonstrated in Paper III, *if relativistic star clusters do form in Nature, they are likely to be very centrally condensed. Moreover, the most interesting astrophysical scenarios—those that may be relevant to the birth of quasars and AGNs—involve the catastrophic collapse of extreme core–halo configurations with relativistic cores and extensive Newtonian halos.* These are precisely the clusters for which one would like to determine the final size of the black hole, but for which our original integrations, based on the relativistic equations given previously, had to terminate well before the evolution was completed.

Equally relevant, numerical problems encountered in evolving extreme core–halo *spherical* clusters often resemble problems that arise when trying to follow the propagation of gravitational waves during the evolution of *nonspherical* systems in general relativity. In the latter situation, one must follow accurately the motion of matter near a black hole event horizon while simultaneously tracking the propagation of any radiation out to much larger radii. The complexity of this problem is also due, in part, to the vast dynamic range that characterizes a space–time with both a localized region of strong gravitational fields as well as a distant region with outgoing gravitational waves. Experience gained in solving the centrally condensed, spherical cluster problem may thus prove useful in the construction of reliable 2 + 1- and 3 + 1-dimensional numerical codes that can handle strong-field space–times with gravitational radiation.

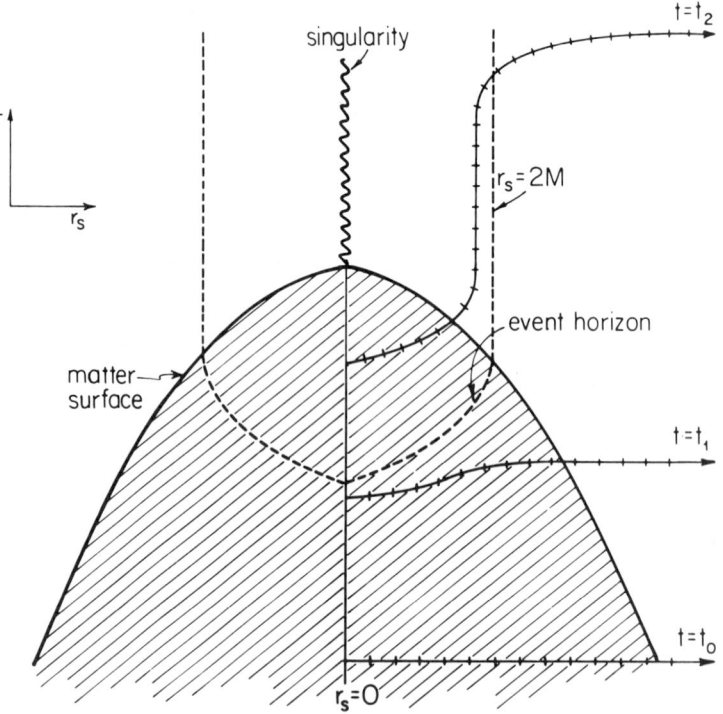

FIGURE 2. Schematic space–time diagram of the collapse of a star cluster to a black hole. The *vertical axis* is the proper time τ of particle worldlines, and the *horizontal axis* is areal (Schwarzschild) coordinate r_s. A singularity forms at the center in a finite proper time. In a numerical calculation one chooses a time slicing t that does not encounter the singularity by slowing down the advance of proper time near the center. Shown are three time slices during the collapse. The *tick marks* on each slice represent a radial grid based on the isotropic coordinate r. When the spatial metric on the time slice t has the "radial" form $A_s^2 dr_s^2 + r_s^2 d\Omega^2$, a spike develops in A_s near the event horizon, as in the usual static Schwarzschild geometry, even though no *physical* singularity is encountered. When the isotropic spatial metric $A^2 dr^2 + A^2 r^2 d\Omega^2$ is used, no spike appears. However, considerable radial grid must be expended along the throat near the horizon in order to accurately determine the metric there.

FIGURE 2 illustrates the main effect responsible for our inability to track the late evolution of highly condensed clusters. Isotropic coordinates, while preventing "spikes" from forming in the radial metric component,[14] lead to considerable *grid stretching* all along the black hole throat. Grid stretching occurs because the isotropic radial coordinate r is forced to span many decades along the black hole throat. The metric field also varies rapidly on the throat. As a consequence, it is necessary to cover the throat with a growing number of grid points to determine the metric accurately in a numerical calculation. However, with the computational limitation of only a finite number of radial grid points available to cover the throat, the growing numerical inaccuracies induced by grid stretching ultimately force the integrations to terminate. Furthermore, maximal time slicing, while successful in holding back the advance of proper time at the center and thus postponing the formation of singularities, causes the lapse function α to decay and the spatial conformal factor A to increase exponentially rapidly at late times at the center. Thus underflows and overflows in the metric components may eventually accompany black hole formation, terminating the integrations before the exterior space–time reaches a final stationary state.

The unsatisfactory situation just described compels us to search for a better choice of space–time coordinates to solve the problem of centrally condensed spherical clusters.

Recall the physical meaning of the lapse function α: it gives the lapse of proper time $d\tau$ measured by an observer at rest in the t = constant hypersurfaces, for a given lapse of coordinate time dt:

$$d\tau = \alpha\, dt. \tag{6.1}$$

Such an observer is called a *normal* observer because his wordline is orthogonal to the t = constant hypersurface.

At late times during gravitational collapse, the lapse goes to zero exponentially with t near the center of the configuration (the "collapse of the lapse"; see Paper I for discussion). We therefore expect quantitites *measured* by the normal observer there to *freeze*, or become constant with t at late times, because there is no lapse of proper time in the normal observer's reference frame.

As an example, consider the radial component of the velocity of a particle, $v^{\hat{r}}$, where the caret denotes a component measured by the normal observer. We have

$$v^{\hat{r}} = \frac{u^{\hat{r}}}{u^{\hat{0}}} = -\frac{\mathbf{u}\cdot\mathbf{e}_{\hat{r}}}{\mathbf{u}\cdot\mathbf{e}_{\hat{0}}}. \tag{6.2}$$

The time basis vector $\mathbf{e}_{\hat{0}}$ is just the 4-velocity of the normal observer, which is the unit normal $\mathbf{n}$ with components given by

$$n_\mu = (-\alpha, 0, 0, 0) \qquad n^\mu = \frac{1}{\alpha}(1, -\beta, 0, 0). \tag{6.3}$$

The radial basis vector is related to the coordinate radial basis vector by $\mathbf{e}_{\hat{r}} = \mathbf{e}_r/A$. Thus equation 6.2 becomes

$$v^{\hat{r}} = \frac{u_r/A}{\alpha u^0}. \tag{6.4}$$

Similarly

$$v^{\hat{\phi}} = \frac{u_\phi/Ar}{\alpha u^0}. \tag{6.5}$$

Substituting equations 6.4 and 6.5 in equation 5.9 gives

$$\alpha u^0 = \frac{1}{[1 - (v^{\hat{r}})^2 - (v^{\hat{\phi}})^2]^{1/2}}. \tag{6.6}$$

Since $v^{\hat{r}}$ and $v^{\hat{\phi}}$ must freeze at late times, so must αu^0, and since u_ϕ is a constant of the motion, we see from equations 6.4 and 6.5 that the *areal* radius of a particle $r_s \equiv Ar$ and the quantity u_r/A also freeze.

It is straightforward to determine which quantities freeze and which do not as $\alpha \rightarrow 0$ (see Paper IV). In particular, as functions of r_s the source profiles ρ and T freeze, while t_r and S_{rr} do not. The quantities β, A, and r (*isotropic* radius of a particle) do not freeze. As a result, equation 5.7 shows that even when $\alpha \rightarrow 0$, the shift β continues to drive changes in r. This is the cause of grid stretching.

EQUATIONS IN ASYMPTOTICALLY FREEZING VARIABLES

To overcome grid stretching and take advantage of α-freezing, we recast the equations in terms of the freezing variables. In either time slicing, the equations of motion of a particle become (Paper IV)

$$\frac{dr_s}{dt} = -\frac{1}{2}\alpha r_s K_T + \alpha \frac{u_r/A}{\alpha u^0}\bigg/\left(1 - \frac{r_s A_{,r_s}}{A}\right). \tag{7.1}$$

$$\frac{d}{dt}\left(\frac{u_r}{A}\right) = \left[-(\alpha u^0)\alpha_{,r_s} + \alpha\frac{u_\phi^2}{r_s^3(\alpha u^0)}\right]\bigg/\left(1 - \frac{r_s A_{,r_s}}{A}\right) + \alpha K_r^r\left(\frac{u_r}{A}\right). \tag{7.2}$$

Since every term on the right-hand sides of equations 7.1 and 7.2 contains an explicit α or $\alpha_{,r}$, we see that

$$\frac{dr_s}{dt} \rightarrow 0 \qquad \frac{d}{dt}\left(\frac{u_r}{A}\right) \rightarrow 0 \quad \text{as} \quad \alpha \rightarrow 0. \tag{7.3}$$

This is a formal demonstration of the freezing of the particle motion at late times near the center of a collapsing configuration in maximal or polar slicing.

In terms of the freezing variables, the metric equation 5.14 is replaced by

$$\left.\frac{\partial A}{\partial t}\right|_{r_s} = \frac{A^2\beta}{r_s} - \frac{1}{2}\alpha A K_T\left(1 - \frac{r_s A_{,r_s}}{A}\right). \tag{7.4}$$

Equation 5.19 for the lapse in polar slicing becomes

$$\alpha = \frac{A_{\max}\alpha_{\max}}{A}\exp\left\{\frac{1}{2}\int_{r_{s,\max}}^{r_s}\left[r_s\left(\frac{A_{,r_s}}{A}\right)^2 + \frac{8\pi r_s S_{rr}}{A^2}\left(1 - \frac{r_s A_{,r_s}}{A}\right)^2\right]dr_s\right\}. \tag{7.5}$$

Equation 5.18 for the lapse in maximal slicing can be similarly transformed to yield

$$\frac{\partial}{\partial r_s}\left(\frac{r_s^2}{1 - r_s A_{,r_s}/A}\frac{\partial \alpha}{\partial r_s}\right) = \alpha r_s^2\left(1 - \frac{r_s A_{,r_s}}{A}\right)\left(\frac{3}{2}K_T^2 + 8\pi\rho + 4\pi T\right). \tag{7.6}$$

Similarly one can transform equation 5.16 for β and 5.17 for K_r^r to equations involving r_s.

THE CODE

Alternative Computational Schemes

The fundamental equations just described can be assembled to yield several distinct computational schemes. The *standard* version uses the *isotropic* radial coordinate r. The *freezing* version uses the *areal* radial coordinate r_s as described in the two preceding sections. Both versions can employ either maximal or polar time slicing.

Our experience indicates that for typical configurations with moderate central concentration the standard code provides the most accurate numerical space–times. However, for extreme configurations with appreciable central concentrations, highly relativistic cores, and extensive Newtonian halos, the freezing version proves to be essential. An enormous dynamic range is necessary to track the evolution of the more distant particles until their fates have been determined, while continuing to follow the central particles accurately. By automatically freezing the coordinates of those particles that wander near or inside a central black hole, this version of the code is able to cope with this dynamic range.

Test-Bed Calculations

We have subjected our code to a rigorous and systematic battery of test-bed calculations. These tests not only served to identify bugs and instabilities, but also to root out inaccuracies and inefficiencies. In addition, they enabled us to calibrate the dynamic range of our code. The problem of dynamic range—the ability to handle problems where the variables span many decades—is particularly important because the applications most often encountered in Nature present exactly this difficulty.

A full discussion of these tests is given in Papers I, II, and IV. As an example, consider the results for Oppenheimer–Snyder collapse. Here we follow the collapse from rest of a homogeneous dust ball (initial radius $R/M = 10$) to a black hole. This is the familiar Friedmann solution, which is known in closed analytic form in Gaussian normal coordinates. For numerical comparisons, one must have available this solution in our coordinate system. The necessary transformations for both maximal and polar slicing have been carried out by Petrich et al.[15,16]

In FIGURE 3 we plot the lapse profile on selected maximal time slices during the collapse. The agreement between the exact and numerical solutions is very good, even after the black hole forms at $t/M \approx 40$.

In FIGURE 4 we plot a space–time diagram for the same case. The matter worldlines are accurately tracked by the numerical code. Moreover, the horizon correctly appears at the origin, grows monotonically outwards, and remains stationary at $r_s/M = 2$ once the last particle crosses inside. Note that maximal time slicing manages to hold back the collapse and prevent the formation of a central singularity. However, it does not do so before the matter surface collapses well inside the horizon,

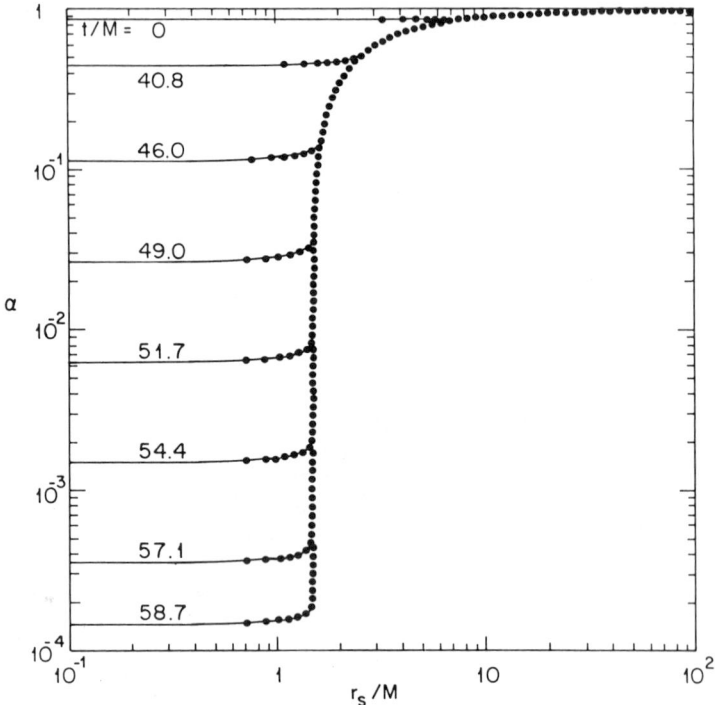

FIGURE 3. The lapse function α as a function of areal radius r_s on selected maximal time slices for Oppenheimer–Snyder collapse from $R = 10M$. The *solid lines* are the results of exact integrations.[15] The *dots* are the results obtained with the numerical code.

to $r_s/M \approx 1.5$, and a region of trapped surfaces forms. This is in contrast to the situation shown in FIGURE 5 for polar slicing. There the matter surface asymptotes to $r_s/M = 2$, and no trapped surface forms. This example nicely illustrates the stronger singularity avoidance property of polar slicing.

THE STABILITY OF RELATIVISTIC STAR CLUSTERS

As one important application of our code, consider the establishment of *binding energy criteria* for the stability of star clusters in general relativity. Imagine a sequence of equilibrium clusters each constructed from the same distribution function, but differing in central redshift z_c. Plot the fractional binding energy E_b/M_0 as a function of z_c along the sequence. Such a plot is shown in FIGURE 6 for relativistic polytropes of index $n = 4$. Unlike the corresponding Newtonian curve, which is monotonic, the relativistic binding energy curve has a *turning point* at sufficiently high redshift, $z_c \approx 0.5$. It is known that for a sequence of *fluid* equilibriums, such a turning point signals the onset of radial instability. No such general theorem has been proved for *collisionless* equilibriums (see references 5, 6, 8, and references therein, for a discussion of previous work). We know from linear perturbation theory (utilizing trial functions in a variational principle) that at very high redshift the configurations are unstable. For some distribution functions, the point of instability appears to coincide with the binding energy maximum, to within numerical accuracy. However, as is clear from the figure, the onset of instability as determined by linear perturbation theory ($\omega^2 < 0$) occurs well beyond the turning point.[7] We do not know whether improved trial functions would show that instability actually sets in earlier along the sequence.

To find out what actually happens, we took models along the equilibrium sequence as initial data for our numerical code. Any numerical inaccuracies in the initial data or the evolution, however small, will induce collapse in an unstable configuration. As is evident from the figure, we find that all configurations beyond the first turning point are dynamically unstable. We thus conclude from this and other examples that *the turning point on the binding energy curve does in fact signify the onset of dynamical instability*. We therefore have "discovered" a theorem awaiting a more formal proof.

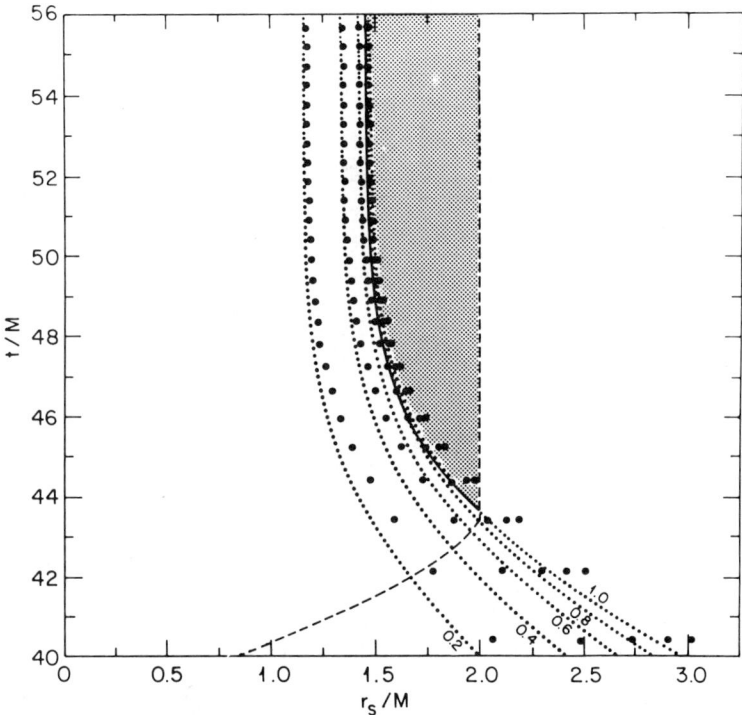

FIGURE 4. Space–time diagram for Oppenheimer–Snyder collapse in maximal slicing from $R = 10M$. The *dotted lines* are the worldlines of Lagrangian matter elements from exact integrations.[15] Each worldline is labeled by the fixed interior rest-mass fraction. The *dots* are points for the corresponding matter elements obtained with the numerical code. The *dashed line* is the event horizon. The *shaded area* is the region of trapped surfaces. Its outer boundary, the apparent horizon, coincides with the event horizon. Its inner boundary is just inside the surface of the matter.

Incidentally, the first unstable model shown in FIGURE 6, with $z_c \approx 0.5$, is of particular interest. It is a prototype of an extreme core–halo configuration, containing only 0.5% of its rest mass in a tiny relativistic core, and the remainder of the mass in an extensive Newtonian halo, extending out to $R/M \approx 5000$. The ratio of the mean to central density is $\langle \rho \rangle / \rho_c = 4.0 \times 10^{-13}$. When this model collapses, a mass much larger than that of the core, approximately 5% of the rest mass, forms a central black hole. At the end of the collapse, the cluster settles in to a new stationary state consisting of a massive Newtonian halo in orbit about a central black hole. *This numerical example provides a viable scenario for the formation of a supermassive black hole in a dense galactic nucleus.*

To handle this case we had to employ the freezing version of our code, which was specially designed for configurations with such large dynamic range. Maximal slicing was not adequate to hold back the collapse. Before the cluster could settle into a final stationary state, the central lapse collapsed below $\alpha_c \approx 10^{-70}$ and the central radial metric coefficient increased above $A_c \approx 10^{49}$. While α could be cut off below some

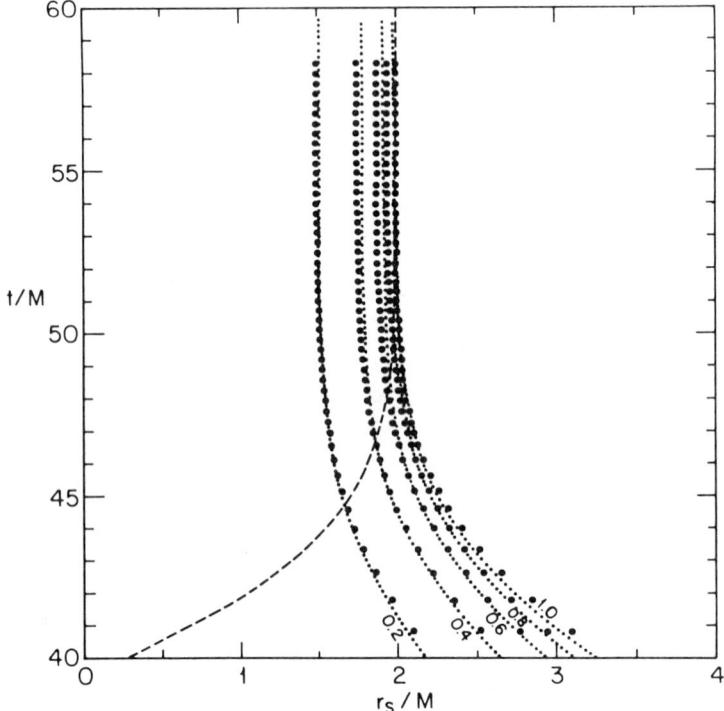

FIGURE 5. Space–time diagram for Oppenheimer–Snyder collapse in polar slicing from $R = 10M$. The *dotted lines* are the worldlines of Lagrangian matter elements from exact integrations.[16] Each worldline is labeled by the fixed interior rest-mass fraction. The *dots* are points for the corresponding matter elements obtained with the numerical code. The *dashed line* is the event horizon.

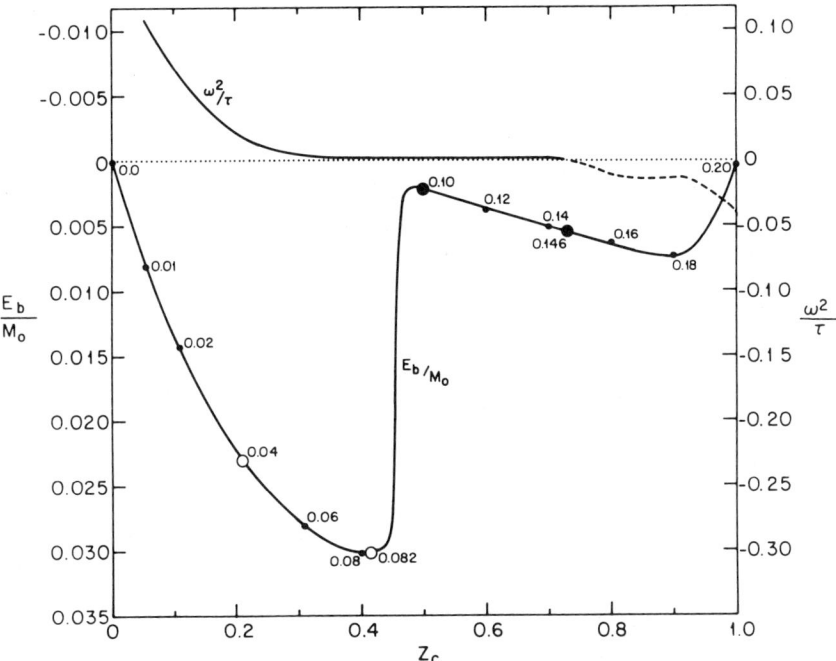

FIGURE 6. The $n = 4$ polytropic sequence. The fractional binding energy E_b/M_0 and oscillation frequency ω^2 in units of $\tau = \rho_c - P_c$ are plotted as functions of central redshift z_c. Clusters along the sequence are labeled by their value of the relativity parameter[7] α. *Open circles* denote stable equilibrium, while *filled circles* indicate collapse to a black hole.

minimum value $\alpha_{min} \ll 1$ to avoid computer underflows, there is no way to cut off the growth of A and still maintain accuracy. Polar slicing in the freezing version proved adequate for this case, since A_c grows much more slowly with time in this gauge.

Several features of the collapse are highlighted in FIGURES 7–10. The initial density profile is shown in FIGURE 7. Snapshots of the central regions of the cluster inside $r_s/M = 2$ at selected times during the collapse are shown in FIGURE 8. The cluster profile does not evolve significantly in the outer Newtonian halo where most of the mass resides. However, the black hole is effective in "sweeping clean" the innermost regions in and around the core. The cluster does not evolve much after $t/M = 40$ except for the steady orbiting of ambient stars about the central hole. The space–time diagram for the collapse is shown in FIGURE 9.

The orbital trajectories of four typical particles near the cluster center are plotted in FIGURE 10. Interior to particle N at $t/M = 0$ resides a fraction $N/7200$ of the total cluster rest-mass. Particle 333, which is initially in an elliptical-like orbit near $r_s/M = 1$, experiences inward spiral motion leading to capture by the black hole after two orbital periods. Particle 410 falls nearly radially from about $7M$ into the black hole. Particle 340 moves nearly unperturbed in a circular orbit at $1M$. Particle 338 moves in a nearly elliptical orbit extending out to $1.5M$ and exhibiting large perihelion precession about the central hole. Pericenter for this orbit appears to be one of the closest of all the ambient particles that do not get captured. It is located at $r_s/M = 0.25$ or $r_s/M_H \approx 5$. This is consistent with the fact that particles that orbit a stationary Schwarzschild black hole inside $r_s/M = 4$ are inevitably captured (see, e.g., reference 17). It is also satisfying that the pericenter position of this marginally stable orbit remains stationary with time, further confirming that by the time our integrations terminate, the cluster has achieved a new dynamic equilibrium about a stationary central black hole.

THE ORIGIN OF QUASARS AND AGNs

There is appreciable evidence, albeit circumstantial, that supermassive black holes with masses in the range $10^6 \lesssim M/M_\odot \lesssim 10^9$ power quasars and AGNs (see reviews by Rees[18] and Begelman et al.[19] and references therein for compelling arguments). The

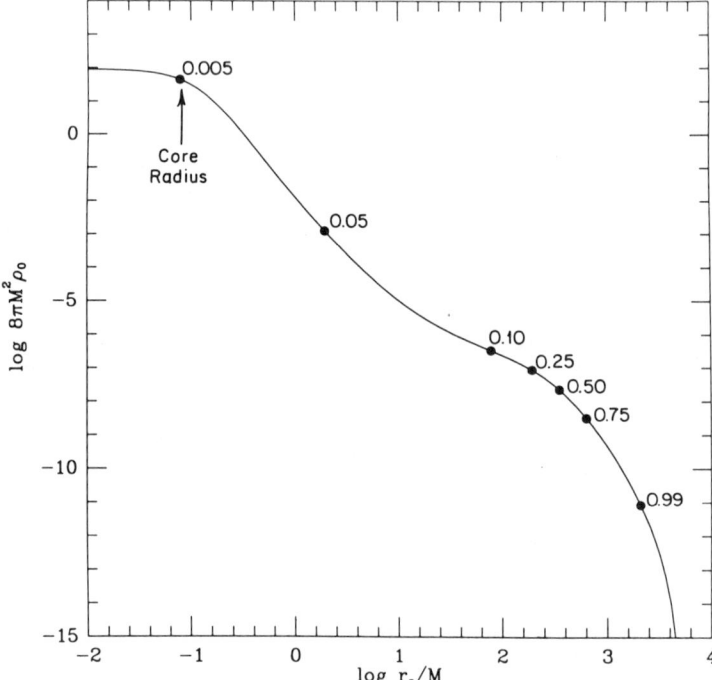

FIGURE 7. The initial rest-mass density profile ρ_0 for the relativistic polytrope with $n = 4$, $\Gamma = 5/4$, and central redshift $z_c = 0.50$. The *dots* are located at points at which the interior rest-mass fraction has the values shown. This is an extreme core–halo configuration, with a highly relativistic core and an extensive Newtonian halo. At the end of the evolution calculation the black hole contains a fraction 0.05 of the total rest mass, roughly 10 times the initial core mass.

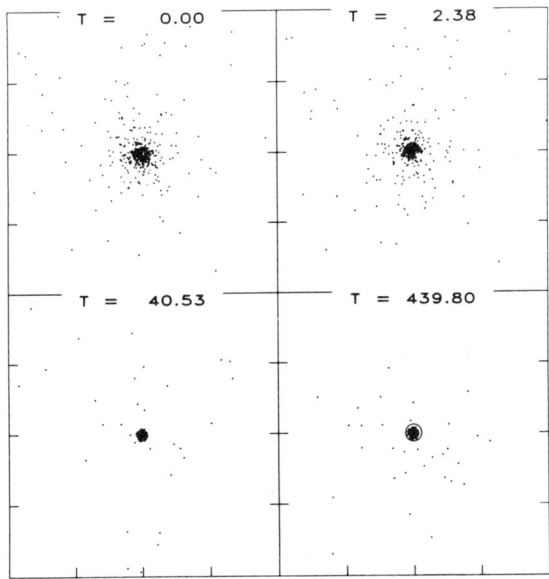

FIGURE 8. Snapshots of the central regions $r_s/M \leq 2$ during the collapse of the extreme core–halo configuration shown in FIGURE 7. The interval between radial grid markers is $\Delta r_s = 1M$. Here the collapse of the innermost regions to a black hole is evident. The *circle* in the last frame shows the event horizon at $r_s/M = 0.1$. Note that the cluster does not evolve appreciably after $t/M = 40$, but is characterized by the steady orbiting of stars about the central black hole.

existence of supermassive black holes may not be restricted to such exotic objects. Our neighbors, like the Andromeda galaxy, probably have black holes of this size at their centers.[20,21] In fact, even our own galaxy may contain a 10^6 $M_\odot$ black hole at the center.[22] Even if black holes are indeed present, a key unanswered question remains: *How and under what circumstances did such supermassive black holes form?*

We proposed in Paper III that the collapse of dense star clusters to supermassive black holes may provide the answer. Our scenario, which is similar to the original proposal of Zel'dovich and Podurets,[9] begins with a dense, but otherwise Newtonian, cluster of compact stars—neutron stars or stellar mass black holes—residing at the center of a galactic nucleus. Because of Coulomb scattering between the stars, the core of such a Newtonian cluster inevitably undergoes a secular collapse on a *relaxation* timescale to a high-density, high redshift state. Such secular Newtonian collapse in a self-gravitating, large N-body stellar system is known as the *gravothermal catastrophe*, or simply *core collapse*. If core collapse should proceed all the way to a relativistic state, then our Vlasov simulations show that the resulting extreme core–halo configuration will be unstable to catastrophic collapse to a black hole. The collapse will take place rapidly, on a *dynamical* timescale. Our calculations show that those clusters capable of reaching such relativistic central redshifts in less than a Hubble time are likely to form supermassive black holes of precisely the right size to explain quasars and AGNs, $10^6 \lesssim M/M_\odot \lesssim 10^9$.

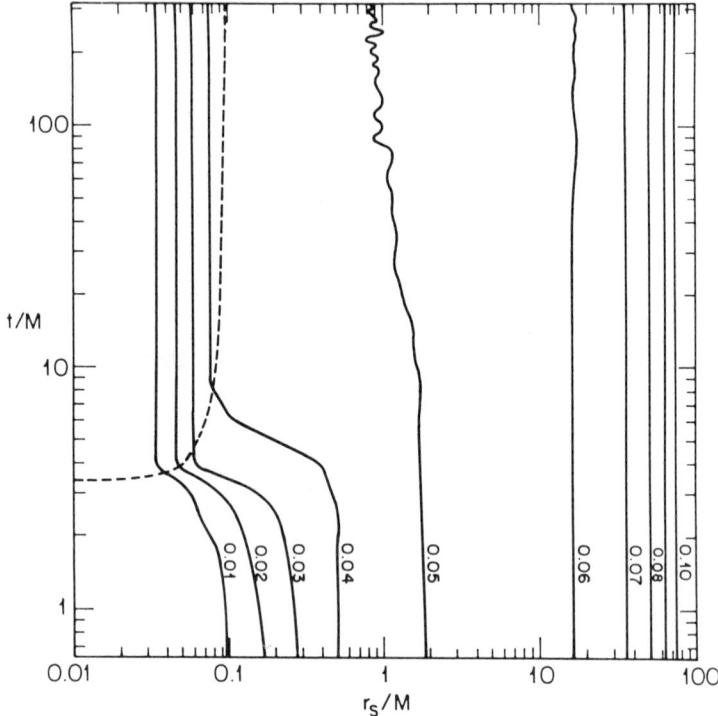

FIGURE 9. Space–time diagram in *polar* slicing for the relativistic polytrope shown in FIGURE 8. The *solid lines* are the worldlines of imaginary Lagrangian matter traces labeled by their fixed interior rest-mass fraction. The *dashed line* is the event horizon. The event horizon asymptotes to $r_s/M = 0.1$, at which point it encompasses 5% of the total cluster rest mass.

Our scenario requires that the initial clusters have at least 10^8 stars moving with velocities in excess of 500 km s^{-1} in the innermost 0.1 pc of a galactic nucleus. Observations of such systems are limited because of the high angular resolution required to identify such compact clusters. It is nevertheless intriguing that the high-resolution balloon-borne telescope abroad *Stratoscope II* did reveal extremely dense stellar systems in galactic nuclei with conditions close to those envisioned here.[23,24] Even the dense core of M31, our nearest neighbor, appears to be ripe for such catastrophic evolution.

Ultimately, planned high-resolution Space Telescope observations of galactic nuclei may determine whether stellar conditions there are indeed suitable for triggering the black hole formation scenario we have proposed.

LESSONS FOR NUMERICAL RELATIVITY

We have succeeded in developing a reliable method that follows the dynamical evolution of collisionless configurations in general relativity. The method is rugged and the space–times generated on the computer are accurate. Though restricted to spherical symmetry, our computational algorithms and numerical examples teach some general lessons about building space–times on the computer. For example, our calculations confirm previous speculations (cf. Piran[25] and references therein) that polar time-slicing has somewhat stronger singularity avoidance features than maximal time-slicing. Accordingly, it is often more desirable to work in the polar gauge for solving certain problems, particularly those for which the dynamical evolution must be followed for a very long time $\Delta t \gg M$ after the appearance of an event horizon to reach a final stationary state.

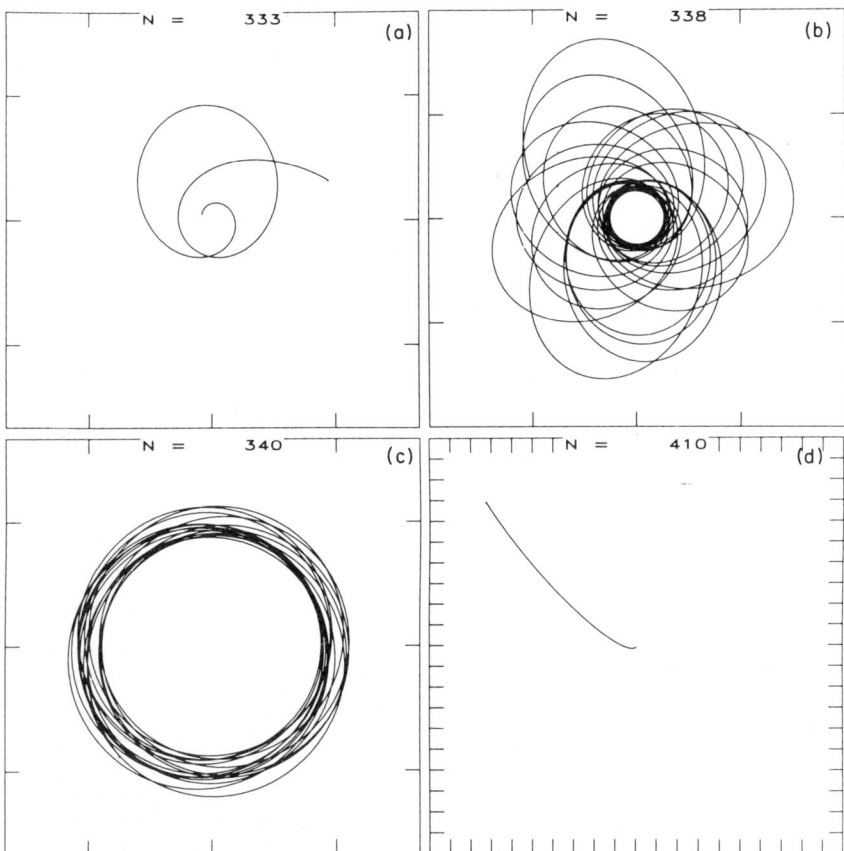

FIGURE 10. Orbital trajectories of 4 typical particles near the cluster center. Each particle N surrounds a fraction $N/7200$ of the total cluster rest mass at $t = 0$. In each frame the spacing between radial grid marks is $\Delta r_s = 1M$. In (**a**) the particle is initially in an elliptical-like orbit, but spirals in to the black hole after roughly two orbits. In (**b**) the particle moves in a nearly elliptical orbit, exhibiting large perihelion precession about the central black hole. Pericenter for this particle, $r_s/M = 0.25$, is one of the closest of all the particles that do not get captured, and remains stationary in time. In (**c**) the particle moves essentially unperturbed in a nearly circular orbit. In (**d**) the particle falls nearly radially into the black hole.

We have also demonstrated, however, that choosing polar slicing is not always in itself sufficient to guarantee numerical success; the selection of well-behaved spatial coordinates is equally crucial to the reliable and trouble-free integration of Einstein's equations on the computer. For example, the adoption of the radial gauge results in the appearance of spikes in the radial metric coefficient whenever a black hole forms. Similarly, the adoption of the isotropic gauge results in severe grid stretching all along a black hole throat.

To circumvent these difficulties we showed that it is sometimes necessary to find variables in terms of which the matter fields effectively freeze near and inside a black hole at late times. We demonstrated that the *areal* radius r_s was such a coordinate and that the ADM equations in *isotropic* coordinates could be easily rewritten in terms of this coordinate in place of the usual isotropic radius r. We suspect that similar coordinates must be employed in higher 2 + 1- and 3 + 1-dimensional codes before really reliable numerical space–times involving black holes can be generated. This suspicion is based on the expectation that in order to accurately evolve gravitational wave data out to large radii, integrations must still continue for a long time $\Delta t \gg M$ after an event horizon forms near the center of the grid. This situation is quite analogous to that already encountered here in spherical symmetry when we considered the collapse of extreme core–halo configurations to black holes. Finding appropriate freezing variables was motivated and greatly simplified by elementary geometric considerations in the spherical case. Discovering analogous variables in more general, nonspherical cases may prove to be more difficult.

Finally, our experience to date in spherical symmetry proves that working with the *collisionless* fluid equations in general relativity by a mean-field particle simulation scheme can be very fruitful. Further exploration of this problem should provide new insights for doing numerical relativity accurately in more complicated, multidimensional cases. We are currently extending our methods to handle such cases.

SUMMARY

We have constructed a new numerical code that solves Einstein's equations for the dynamical evolution of a *collisionless* gas of particles in general relativity. Our initial investigation was restricted to spherically symmetric systems, but the gravitational field can be arbitrarily strong and particle velocities can be arbitrarily close to the speed of light. Our computational scheme combined the tools of numerical relativity with those of N-body particle simulation. We solved the Vlasov equation in general relativity by particle simulation and determined the gravitational field using the ADM 3 + 1 formalism. Physical applications included the stability of relativistic star clusters, the binding energy criterion for stability, the collapse of star clusters to black holes, and relativistic violent relaxation. Astrophysical applications included the possible origin of quasars and active galactic nuclei via the collapse of dense star clusters to supermassive black holes. We found that our method is extremely accurate, even in the case of black hole formation. It provides a unique proving ground for testing different computational algorithms and gauge choices for the construction of numerical space–times.

REFERENCES

1. SHAPIRO, S. L. & S. A. TEUKOLSKY. 1985. Astrophys. J. **298:** 34 (Paper I).
2. SHAPIRO, S. L. & S. A. TEUKOLSKY. 1985. Astrophys. J. **298:** 58 (Paper II).
3. SHAPIRO, S. L. & S. A. TEUKOLSKY. 1985. Astrophys. J. (Lett.) **292:** L41 (Paper III).
4. SHAPIRO, S. L. & S. A. TEUKOLSKY. 1986. Astrophys. J. **307:** 575 (Paper IV).
5. IPSER, J. R. & K. S. THORNE. 1968. Astrophys. J. **154:** 251.
6. IPSER, J. R. 1969. Astrophys. J. **158:** 17.
7. FACKERELL, E. D. 1970. Astrophys. J. **160:** 859.
8. IPSER, J. R. 1980. Astrophys. J. **238:** 1101.
9. ZEL'DOVICH, Ya. B. & M. A. PODURETS. 1965. Astron. Zh. **42:** 963. (English transl. in Sov. Astron.—AJ **9:** 742.)
10. ZEL'DOVICH, Ya. B. & I. D. NOVIKOV. 1971. Relativistic Astrophysics. Vol. 1. Univ. of Chicago Press. Chicago.
11. ARNOWITT, R., DESER, S. & C. W. MISNER. 1962. *In* Gravitation: An Introduction to Current Research. L. Witten, Ed. Wiley. New York.
12. EARDLEY, D. M. 1982. Unpublished talk at Numerical Astrophysics: A Symposium in Honor of James R. Wilson. Univ. of Illinois, Urbana, Oct. 1982.
13. BARDEEN, J. M. & T. PIRAN. 1983. Phys. Rep. **96:** 205.
14. SMARR, L. & J. W. YORK. 1978. Phys. Rev. D **17:** 1945.
15. PETRICH, L., S. L. SHAPIRO & S. A. TEUKOLSKY. 1985. Phys. Rev. D **31:** 2459.
16. PETRICH, L., S. L. SHAPIRO & S. A. TEUKOLSKY. 1985. Phys. Rev. D **33:** 2100.
17. MISNER, C. W., K. S. THORNE & J. A. WHEELER. 1973. Gravitation. Freeman. San Francisco.
18. REES, M. J. 1984. Ann. Rev. Astron. Astrophys. **22:** 471.
19. BEGELMAN, M. C., R. D. BLANDFORD & M. J. REES. 1984. Rev. Mod. Phys. **56:** 255.
20. KORMENDY, J. 1988. Astrophys. J. **325.** In press.
21. DRESSLER, A. & D. O. RICHSTONE. 1988. Astrophys. J. **324.** In press.
22. LACY, J. H., F. BAAS, C. H. TOWNES & T. R. GEBALLE. 1979. Astrophys. J. (Lett.) **227:** L17.
23. SCHWARZSCHILD, M. 1973. Astrophys. J. **182:** 357.
24. LIGHT, E. S., R. E. DANIELSON & M. SCHWARZSCHILD. 1974. Astrophys. J. **194:** 257.
25. PIRAN, T. 1983. *In* Gravitational Radiation, Les Houches 1982. N. Deruelle and T. Piran, Eds.: 203. North-Holland. Amsterdam.

Repulsive and Attractive Double-Bubble Space-Times

JAMES R. IPSER

Department of Physics
University of Florida
Gainesville, Florida 32611

INTRODUCTION

It is a well-known fact that the collection of physically reasonable, integrable (i.e., exact) dynamical solutions for stress-energy sources in general relativity is sparse. Few examples exist besides the well-known Friedmann–Robertson–Walker cosmologies and the solutions for collapsing homogeneous (Oppenheimer–Snyder) dust balls and spherical dust shells.

Here we shall discuss some integrable cosmological solutions obtained recently,[1-3] involving thin domain walls exhibiting surface energy and wall tension or pressure. Such structures might arise as topological defects associated with phase transitions in the early universe, as has been suggested by various particle physics theories. We shall focus our attention on solutions for which the universe consists of two bubblelike regions (i.e., interiors of two spheres) glued together at their common boundary—a spherical wall that surrounds and encloses each of them.

THE BASIC EQUATIONS

We focus on space–time with a three-dimensional timelike hypersurface S that contains stress energy. With ξ^a the unit spacelike normal ($\xi_a \xi^a = +1$) to the surface, the three-metric intrinsic to the hypersurface is

$$h_{ab} = g_{ab} - \xi_a \xi_b, \tag{2.1}$$

where g_{ab} is the four-metric of space–time. In terms of the covariant derivative ∇_a associated with g_{ab}, the extrinsic curvature π_{ab} of S is given by

$$\pi_{ab} = D_a \xi_b \qquad D_a = h_a^b \nabla_b. \tag{2.2}$$

At the wall, the components of Einstein's equation with one or both indices parallel to ξ^a take the form[1]

$$^3R + \pi_{ab}\pi^{ab} - \pi^2 = -16\pi G_N T_{ab} \xi^a \xi^b, \tag{2.3a}$$

$$h_{ab}D_c\pi^{bc} - D_a\pi = 8\pi G_N T_{bc} h_a^b \xi^c, \tag{2.3b}$$

where T^{ab} is the stress-energy tensor off the wall and πG_N is the geometrical constant 3.14 ... times Newton's constant. The remaining components of Einstein's equation determine the surface stress-energy tensor S_{ab} of S:

$$S_{ab} = \int dl (T_{ab})_{\text{wall}} = -\frac{1}{8\pi G_N}(\gamma_{ab} - h_{ab}\gamma_c^c), \tag{2.4}$$

where

$$\gamma_{ab} \equiv \pi_{+ab} - \pi_{-ab}. \tag{2.5}$$

Here l is the proper distance through S in the direction of the normal ξ^a, and the subscripts $\pm$ refer to values just off S on the side determined by the direction of $\pm\xi^a$. It will be convenient to use the general notation

$$[Q]_-^+ \equiv Q_+ - Q_- \qquad \{Q\}_-^+ \equiv Q_+ + Q_- \tag{2.6}$$

for any quantity Q. Hence, $\gamma_{ab} = [\pi_{ab}]_-^+$.

We now follow the usual procedure[1,2] of taking the sums and differences of equations 2.3 on opposite sides of S. This yields

$$h_{ce}D_b S^{eb} = -[T_{ab}h_c^a\xi^b]_-^+, \tag{2.7a}$$

$$h_{ce}D_b\{\pi^{eb}\}_-^+ - D_c\{\pi_a^a\}_-^+ = 8\pi G_N\{T_{ab}h_c^a\xi^b\}_-^+, \tag{2.7b}$$

$$\{\pi_{ab}\}_-^+ s^{ab} = 2[T_{ab}\xi^a\xi^b]_-^+, \tag{2.7c}$$

$$^3R + \tfrac{1}{4}[\{\pi_{ab}\}_-^+\{\pi^{ab}\}_-^+ - (\{\pi_a^a\}_-^+)^2]$$
$$= -16\pi^2 G_N^2[S_{ab}S^{ab} - \tfrac{1}{2}(S_a^a)^2] - 8\pi G_N\{T_{ab}\xi^a\xi^b\}_-^+. \tag{2.7d}$$

Here ^3R is the Ricci scalar curvature of h_{ab}.

Next we examine the acceleration of observers comoving with S. Let u^a be the 4-velocity of an observer on S who sees no energy flux, and extend u^a off S in smooth fashion to obtain the 4-velocity of hovering observers. Thus obtain

$$u^a\nabla_a u^e = h_b^e u^a\nabla_a u^b - \xi^e u^a u^b \pi_{ab}, \tag{2.8}$$

so that

$$\{\xi_e u^a\nabla_a u^e\}_-^+ = -u^a u^b\{\pi_{ab}\}_-^+ \tag{2.9a}$$

$$[\xi_e u^a\nabla_a u^e]_-^+ = -u^a u^b\gamma_{ab} = 8\pi G_N(S_{ab}u^a u^b + \tfrac{1}{2}S_c^c). \tag{2.9b}$$

Our focus on space–times with identical bubbles leads us to impose the constraint of reflection symmetry,

$$\{\pi^{ab}\}_-^+ = 0, \tag{2.10}$$

which in turn implies, via equations 2.7b, 2.7c, and 2.9a

$$\{T_{ab}h_c^a\xi^b\}_-^+ = [T_{ab}\xi^a\xi^b]_-^+ = \{\xi_e u^a\nabla_a u^e\}_-^+ = 0. \tag{2.11}$$

Further, our interest in walls with surface energy and tension or pressure leads us to take the surface energy tensor to be of the form

$$S^{ab} = \sigma u^a u^b - \tau(h^{ab} + u^a u^b), \quad \frac{\tau}{\sigma} \equiv \Gamma = \text{const.} \leq 1. \tag{2.12}$$

Equations 2.7a and 2.9b now take the form

$$u_c D_b[(\sigma - \tau)u^b] + (\sigma - \tau)h_{cd}u^b D_b u^d - D_c\tau = -[T_{ab}h_c{}^a\xi^b]_-^+, \quad (2.13)$$

$$\xi_e u^a \nabla_a u^e|_+ = -\xi_e u^a \nabla_a u^e|_- = 2\pi G_N(\sigma - 2\tau). \quad (2.14)$$

Once a solution off the wall is specified, the major effect focuses on solving these latter two equations, as we shall next see for two types of cosmological solutions.

SCHWARZSCHILD DOUBLE BUBBLES

Here we take each side off the wall to be a Schwarzschild vacuum with line element

$$(ds^2)_\pm = (-e^\nu dt^2 + e^\lambda dr^2 + r^2 d\Omega^2)_\pm \qquad e^\nu = e^{-\lambda} = (1 - 2G_N M/r). \quad (3.1)$$

Letting $R(t)$ denote the common value of $r_\pm$ at the wall, and taking $t_\pm$ to have the same value at a given event on the wall, we obtain $(a = t, r, \theta, \phi)$

$$u_\pm^a = (\dot{t}, \dot{R}, 0, 0) \qquad \xi_\pm^a = (\mp e^{-\nu}\dot{R}, \mp e^\nu \dot{t}, 0, 0), \quad (3.2)$$

where an overdot denotes derivative with respect to proper time of the comoving observer with 4-velocity u^a. Equations 2.13 and 2.14 now yield

$$-\frac{1}{e^\nu \dot{t}}\left(\ddot{R} + \frac{G_N M}{R^2}\right) = 2\pi G_N(1 - 2\Gamma)\sigma = 2\pi G_N(1 - 2\Gamma)\sigma_0 R^{-2(1-\Gamma)}, \quad (3.3)$$

where σ_0 is a constant. This equation is easily integrated, yielding the first integral

$$\left(1 - \frac{2G_N M}{R} + \dot{R}^2\right)^{1/2} - 2\pi G_N \sigma_0 R^{2\Gamma - 1} = \text{const.} = 0. \quad (3.4)$$

That the constant of integration must vanish is verified by demanding that equation 2.7d be satisfied.

The dynamics of the bubble/wall system, as given by equation 3.4, depends on the value of Γ. For $\Gamma \leq \frac{1}{2}$ the wall, and each bubble, expands to a certain maximum radius and subsequently collapses. The wall is gravitationally attractive in this case. For $\Gamma > \frac{1}{2}$, on the other hand, the wall either expands out to a maximum radius and then collapses or it contracts down to a minimum radius and then expands. This latter behavior reflects the fact that the wall is gravitationally repulsive in this case. In this connection, one easily verifies that for $\Gamma = 1$ (usual domain wall) and $M = 0$ (flat interior on either side of wall), a solution is[1]

$$R^2 = \frac{1}{(2\pi G\sigma_0)^2} + t^2, \quad (3.5)$$

corresponding to a wall that comes in from infinity, reaches a minimum radius, and then expands out to infinity, all the while accelerating away from its interior.

FRIEDMANN DOUBLE BUBBLES

Here we take each side off the wall to be a Friedmann cosmology with line element

$$(ds^2)_\pm = (-dt^2 + a^2(t)[d\chi^2 + f^2(\chi)d\Omega^2])_\pm \tag{4.1}$$

and with perfect-fluid stress-energy tensor

$$T^{ab} = (\epsilon + p)v^a v^b + pg^{ab}. \tag{4.2}$$

We assume that $\chi_\pm = X(t)$ on the wall, so that $(a = t, \chi, \theta, \phi)$

$$v^a = (1, 0, 0, 0) \quad u^a = (\dot{t}, \dot{X}, 0, 0) \quad \xi^a_\pm = (\mp a\dot{X}, \mp \dot{t}/a). \tag{4.3}$$

Equations 2.13 and 2.14 now yield

$$\dot{\sigma} + (1 - \Gamma)\sigma D_a u^a = 2\,(\epsilon + p)v_a v_b u^a \xi^b|_+ = -2(\epsilon + p)a\dot{t}\dot{X}, \tag{4.4}$$

$$-(1/a\dot{t})\,(a^2\dot{X}) = 2\pi G_N(1 - 2\Gamma)\sigma. \tag{4.5}$$

The value $\Gamma = \frac{1}{2}$ corresponds to an isotropic network of cosmic strings.[1] In this case, a particularly simple solution exists, viz.,

$$X = \text{const.} \quad \sigma = \text{const.}/a.$$

Hence, the wall simply comoves in the usual sense, expanding or contracting in unison with the rest of the universe. This reflects the fact that for $\Gamma = \frac{1}{2}$ the wall is neither attractive nor repulsive: observers hover next to it simply by moving along the geodesics of comoving observers in the Friedmann geometry.

Equations 4.4 and 4.5 are not integrable in analytic form for the usual domain-wall case $\Gamma = 1$, but it is clear from their form that the wall is now repulsive, in the sense that an observer must accelerate toward the wall, and away from the interior of his Friedmann bubble, in order to remain next to the wall. In all of these cases, observers in either bubble would be unaware of the wall and the opposite bubble unless they were able to perform observations sensitive to the full global geometry of the universe.

REFERENCES

1. IPSER, J. R. & P. SIKIVIE. 1984. Phys. Rev. D **30**:712–719.
2. IPSER, J. R. 1984. Phys. Rev. D **30**: 2452–2456.
3. IPSER, J. R. 1987. Phys. Rev. D **36**: 1933–1935.

Integrability of Magnetic Confinement Systems

R. V. E. LOVELACE

Department of Applied Physics
Cornell University
Ithaca, New York 14853

The integrability of the single-particle motion is central to our understanding of the equilibrium and stability of laboratory magnetic confinement systems. Two generic types of magnetic confinement systems are used to illustrate the importance of the particle motion. The first system is the axisymmetric field-reversed configuration or Astron system (first proposed by Christofilos[1]) for the magnetic confinement of high-temperature ($T \sim 10^8$ K) fusion plasmas. The second system is the dynamic magnetic trap for the magnetic confinement of low-temperature ($<10^{-3}$ K) spin-polarized hydrogen gas.[2] In the Astron system, an essentially uniform constant external B_z magnetic field is used to support the gyration of a large number, $N \sim 10^{18}$, of energetic ions. Electrostatic fields are neutralized by a background plasma, but there is a net azimuthal current ($J_\phi < 0$) due to the gyratory motion of the energetic particles. The magnetic field is thus the sum of the external field and the self field due to the particle ring. With a sufficiently large number of particles, a closed magnetic field configuration can be achieved where the direction of the total field at the ring's center is opposite to that of the external field. This "reversed field configuration," which has been produced experimentally,[3] is predicted to behave as a magnetic bottle for confining plasma. An important aspect of the Astron system is the dynamics of the energetic particles that are to a first approximation collisionless. An ion may encircle the axis more than 10^6 times between collisions. Because the Astron system is axisymmetric and time-independent, the canonical angular momentum, $P_\phi = mrv_\phi + (q/c)rA_\phi(r, z)$, and the Hamiltonian, $H = [\mathbf{P} - q/c\,\mathbf{A}]^2/(2m)$, are rigorous constants of the motion. Here, $\mathbf{P}$ is the canonical momentum and $\mathbf{A} = A_\phi \hat{\phi}$ is the vector potential. An important question concerns the existence and usefulness of a third constant of the motion. Most of the theoretical studies of Astron equilibria, their adiabatic compression, and their collisional evolution have assumed for theoretical convenience that there is not a third constant of the motion. On the other hand, analytical and numerical studies[4-6] of the single-particle orbits indicate that there often is at least an approximate third constant of the motion. Only in two limiting ring geometries does it appear, however, that an approximate third constant of the motion will be sufficiently simple to be useful for analytic studies of equilibria, compression, etc. For large aspect ratio, "bicycle-tire" ring equilibria, the effective potential for the poloidal (r, z) motion, $[P_\phi - (q/c) \cdot rA_\phi]^2/(2mr^2)$, has circular symmetry in the (r, z) plane about the point of the potential minimum. In this case, the third constant of the motion is an "angular-momentum" in the (r, z) plane.[5] In another extreme, that of axially elongated rings (with axial length much larger than the radius), the third constant of the single particle motion may be the adiabatic invariant $I_r = \oint dr\, p_r$. The particles make many radial bounces for every axial bounce.

The second magnetic confinement system we consider is a proposed "dynamic trap" for the magnetic confinement of a gas of spin-polarized neutral hydrogen gas.[2] The trap uses the interaction of the magnetic moments of the gas atoms with a time-dependent magnetic field. The magnetic moments of a gas of hydrogen atoms can be aligned by well-known experimental techniques.[7] The potential energy of an atom in a nonuniform magnetic field is $V = -\mu \cdot \mathbf{B}$, where μ is the atom's magnetic moment. Atoms with μ antiparallel to $\mathbf{B}$ tend to move toward the minimum $|\mathbf{B}|$, and are referred to as weak field seekers, whereas atoms with μ parallel to $\mathbf{B}$ are strong field seekers. For the antiparallel case, static magnetic field traps have been constructed that have an isolated point where $|\mathbf{B}(r, z)|$ is a minimum. However, the "antiparallel atoms" are known to be much less stable (to spin-flipping and recombination) than the atoms with μ parallel to $\mathbf{B}$. For this reason we have proposed and studied theoretically a confinement system for "parallel atoms." The simplest version of such a system consists of a uniform constant B_z field and the field of a circular current loop of radius R, the current I in which oscillates at an angular frequency ω_e. Near the ring center the single-particle equations of motion are $d^2x/dt^2 = -\alpha(t)x$, $d^2y/dt^2 = -\alpha(t)y$, and $d^2z/dt^2 = +2\alpha(t)z$, where $\alpha(t) = \alpha_0 \sin(\omega_e t)$, $\alpha_0 \equiv 3\pi\mu I_0/(cmR^3)$, and $I(t) = I_0 \cdot \sin(\omega_e t)$. The single-particle motion is confined for $\alpha_0 \leq 0.18(\omega_e)^2$. The particle motion, $\mathbf{x}(t) = \mathbf{X}(t) + \xi(t)$, is seen to consist of a fast, "gyratory" motion, $\xi(t)$, and a slow, "guiding center" motion, $\mathbf{X}(t)$. The guiding center motion is well accounted for by a Hamiltonian, which can be written as $H = \frac{1}{2}(\dot{\mathbf{X}})^2 + V_e(\mathbf{X})$, where $V_e = (\frac{1}{4}) \cdot (\alpha_0/\omega_e)^2 (X_x^2 + X_y^2 + 4X_z^2)$ is an effective potential. This potential is analogous to the pondermotive potential for, say, an electron moving in an electromagnetic wave. The existence of the integral H is important in that it permits a treatment of the high-density regime where collisions are important.[2] Also, this integral permits a simple treatment of adiabatic compression. The classical treatment of dynamic traps has provided a guide for solving the quantum mechanics of a dynamic trap.[8]

REFERENCES

1. CHRISTOFILOS, N. C. 1958. In Proc. 2nd United Nations Int. Conf. on the Peaceful Uses of Atomic Energy, Vol. **32**: 279. United Nations, Geneva, Switzerland.
2. LOVELACE, R. V. E., C. MEHANIAN, T. J. TOMMILA & D. M. LEE. 1985. Nature **318**: 30.
3. FLEISCHMANN, H. H., D. J. REJ, IM. TUSEZEWSKI & S. C. LUCKHARDT. 1979. In Proc. 3rd Int. Conf. on Collective Methods of Acceleration. Harwood Academic, London. 699.
4. FINN, J. M. 1979. Plasma Phys. **21**: 405.
5. LARRABEE, D. A. & R. V. E. LOVELACE. 1980. Phys. Fluids **23**: 1436.
6. BERK, H. L., H. MOMOTA & T. TAJIMA. 1988. Phys. Fluids. In press.
7. SILVERA, I. F. 1982. Physics **109/110B**: 1499.
8. LOVELACE, R. V. E. & T. J. TOMMILA. 1987. Phys. Rev. A **35**: 3597.

Eternity, Chaos, Lie Algebras, Integrability, and Accelerator Design[a]

ALEX J. DRAGT

Department of Physics and Astronomy
University of Maryland
College Park, Maryland 20742

Very large accelerator–storage rings have been and are being constructed to study the fundamental nature of matter and its behavior at very high energies. The Tevatron at Fermilab is 6 kilometers in circumference; the LEP facility near completion at CERN is 27 kilometers in circumference; and the proposed Superconducting Super Collider[1] is planned to be 90 kilometers in circumference. These machines are designed to accelerate charged particles (protons or electrons) to very high energies and to then store them for large periods of time. For example, the Superconducting Super Collider will accelerate protons to an energy of 20 TeV and then store them for at least 8 hours.

From the viewpoint of dynamical systems, these large storage times represent a near eternity. For example, during 8 hours in the Superconducting Super Collider, a proton will make approximately 10^8 turns around the ring, will make approximately 10^{10} betatron oscillations, and will experience approximately 10^{12} magnetic deflections. Thus, in order to ensure good performance, it is essential to have very good long-term orbit stability. That is, particles must actually be stored within the ring and lie on well-confined orbits for long periods of time.

One approach to the study of long-term orbit behavior is through the use of maps. This method represents the action of each separate element of a beamline or accelerator, including nonlinear effects, by a certain transfer map. Each transfer map describes the relation between the initial condition in phase space of a particle entering a particular element and the final condition in phase space of the same particle as it exists that element. These maps can then be combined, following well-defined multiplication rules, to obtain a resultant one-turn transfer map that characterizes the entire system.

The stability of orbits is governed by the behavior of the one-turn transfer map under repeated iteration. That is, suppose one selects a particular point in phase space corresponding to the initial conditions for the trajectory of a particular particle as it begins a turn. Then the phase–space coordinates at the end of this turn are gotten by applying the one-turn transfer map to the initial phase–space point. Similarly, the results of successive turns are gotten by applying the one-turn transfer map over and over again.

If the phase–space points obtained by applying the one-turn transfer map over and over again to a particular initial condition lie on a Kolmogorov-Arnold-Moser (KAM) torus, then particles on the trajectory with that initial condition will have long-term

---

[a]This work was supported in part by the U.S. Department of Energy under Contract AS05-80ER10666.

stability. By contrast, if the succession of phase–space points exhibits chaotic behavior, then particles on this trajectory will generally be lost in a relatively short period of time. It follows that questions of long-term stability are closely related to questions of chaotic behavior. In particular, one wants to know under what conditions there will be, at worst, only very small-scale chaotic behavior.

It is evident that the question of the existence of large-scale chaotic behavior or, alternatively, of KAM tori is closely related to nonlinear properties of the one-turn transfer map. The nonlinearity of this map is in turn related to the magnetic field quality of the magnets used to construct the accelerator–storage ring. If the magnets have very uniform fields, then the one-turn transfer map will be very nearly linear. Correspondingly, there will be very little chaotic behavior. Conversely, if the magnets have nonuniform fields with large high-order multipole (sextupole, octupole, etc.) content, then the one-turn transfer map will be very nonlinear. Correspondingly, there will be large regions of phase space for which orbits exhibit very chaotic behavior, and the performance of the accelerator–storage ring will be unacceptable.

Unfortunately, the construction of magnets with very good field quality is very expensive. Also, magnet cost is a major component in the overall cost of an accelerator–storage ring. For example, the cost of the magnets for the Superconducting Super Collider is expected to be about $1 billion. Consequently, in this situation, there is a very high price tag on the understanding of chaotic behavior and how it can be minimized.

In the context of accelerator design, there are at least two aspects to the study of chaotic behavior. The first aspect is that of detection: Given a set of phase–space points, how does one tell if they lie near or on some torus in a multidimensional (6-dimensional in the case of accelerators) phase space, or instead exhibit chaotic behavior? The second aspect is that of prediction and control: Given a one-turn transfer map, how does one predict whether or not applying it over and over again to a given initial condition will indeed lead to chaotic behavior, and what alterations in the map would lead to less chaotic behavior?

These and related questions are currently being studied with the aid of Lie algebraic methods.[2] It can be shown that the set of transfer maps forms an infinite dimensional Lie group. Moreover, each group element can be described in terms of certain polynomials in the phase–space variables. Thus, the existence or nonexistence of chaotic behavior is in principle related to the sizes of the coefficients in these polynomials.

Also, it can be shown that the one-turn transfer map can, under certain circumstances, be brought to a certain normal form by the nonlinear analog of performing a similarity transformation. If it exists, this normal form map has the property that it maps perfect tori into themselves. (A perfect torus is defined to be the topological product of circles in the pairs of phase–space variables q_i, p_i.) Correspondingly, if it exists, the transforming map used to produce the "similarity" transformation has the property of mapping KAM tori, if they exist, into perfect tori. It is easy to check for the existence of perfect tori by making projections onto the q_i, p_i planes. Thus, Lie algebraic methods appear to provide a natural procedure for the detection of KAM tori or, alternatively, chaotic behavior.

From a slightly different perspective, the existence of KAM tori is related to the existence of integrals of motion. First, it can be shown that the action of transfer maps, which originally was defined only on the set of phase–space variables, can be extended to include a well-defined action on any function of the phase–space variables. Second, it is quite easy to find integrals for a map in normal form. Finally, once these integrals have been found, one can find integrals for the original one-turn transfer map simply by applying the transforming map to the integrals of the normal form map.

The problem of determining the expected amount of chaotic behavior, starting from the size of various polynomial coefficients, is still being investigated. Attention is currently being devoted to the development of special normal forms in the presence of resonances, and to various methods for computing homoclinic separatrix splitting.

In summary, the detection of chaotic behavior, and its limitation and control, play an important role in accelerator design. Lie algebraic methods appear to offer considerable promise in dealing with these questions. In particular, they can be used to detect and construct KAM tori and their related integrals. They may also be useful in predicting when large-scale chaotic behavior will occur.

REFERENCES

1. For a description of the Superconducting Super Collider, see Conceptual Design of the Superconducting Super Collider, Rep. SSC-SR-2020 (1986), a publication of the SSC Central Design Group, c/o Lawrence Berkeley Laboratory, 90/4040, 1 Cyclotron Road, Berkeley, CA 94720.
2. For a description of Lie algebraic methods, see DRAGT, A. J. Elementary and advanced Lie algebraic methods with applications to accelerator design, electron microscopes, and light optics, Nucl. Instrum. Methods Phys. Res. **A258**: 339–354. 1987. See also Dragt, A. J., *et al. 1988*. Lie algebraic treatment of linear and nonlinear beam dynamics. Accepted for publication in Ann. Rev. Nucl. Part. Sci. **38.**

Hannay's Angle and Berry's Phase in the Classical Adiabatic Motion of Charged Particles[a]

ROBERT G. LITTLEJOHN[b]

Center for Nonlinear Studies
Los Alamos National Laboratory
Los Alamos, New Mexico 87545

INTRODUCTION

An anholonomic effect occurs when a physical system is transported through some sequence of states, and when the result of the transport process depends not only on the final state but also on the history or sequence of intermediate states. This paper will deal with phase anholonomies, which can be represented by inexact differentials whose integrals are path dependent. The best known phase anholonomy in physics today is Berry's phase,[1] whose classical analog is Hannay's angle.[2] Since this paper will deal with a classical problem, the analogy with Hannay's angle is particularly strong, and it is appropriate to begin by summarizing the principal features of Hannay's angle. It turns out that the phase anhonomomy that occurs in the adiabatic motion of a charged particle in a strong magnetic field, otherwise known as guiding center motion, has certain features that can be subsumed under the general analysis of Hannay, and other features that go beyond this, such as the transport of frames in 3-dimensional physical space.

Hannay's angle emerges from asking new questions about the old adiabatic theorem. Following Hannay,[2] we consider a classical Hamiltonian $H(q, p, \mathbf{X})$, where $\mathbf{X} = (X_1, \ldots, X_n)$ represents a collection of parameters, considered to be slow functions of time. We also let (θ, I) be the the action-angle variables associated with this Hamiltonian. Naturally, (θ, I) are not only functions of (q, p) but also of $\mathbf{X}$, and therefore of t. Under these circumstances, the usual adiabatic theorem states that the action I is conserved, at least in an averaged sense.

Hannay extends this by asking about the angle θ, and he finds the following answer. The evolution equation for the angle can be written in the form

$$\dot{\theta} = \omega_0(I, \mathbf{X}(t)) + \Delta\dot{\theta}, \qquad (1)$$

where $\omega_0(I, \mathbf{X})$ represents the frequency the system would have if the parameters were "frozen," that is, it is $\partial H/\partial I$ when H has been reexpressed as a function of the action,

---

[a]This work was supported by the U.S. Department of Energy under Contracts DE-AC03-76SF00098 and W7405/ENG-36, and by the National Science Foundation under Grant NSF-PYI-84-51276.
[b]Permanent address: Department of Physics, University of California, Berkeley, California 94720.

and where the second term is the one coming from the time derivative of the generating function. The quantity $\Delta\theta$ is called "Hannay's angle," and Hanny shows that it can be written in the form

$$\Delta\theta = \int_0^t \sum_i A_i(I, \mathbf{X}) \dot{X}_i \, dt = \int_{\mathbf{x}_0}^{\mathbf{x}_1} \mathbf{A}(I, \mathbf{X}) \cdot d\mathbf{X}, \tag{2}$$

where $\mathbf{X}_0$ and $\mathbf{X}_1$ are the initial and final parameter values, and where $\mathbf{A}$ is an n-dimensional vector defined by

$$A_i(I, \mathbf{X}) = \frac{1}{2\pi} \int d\theta \, \frac{\partial \theta}{\partial X_i}(\theta, I, \mathbf{X}). \tag{3}$$

Equation 2 is to be interpreted in an averaged sense.

The striking thing about Hannay's angle is that it is independent of the rate at which the adiabatic process is carried out, so long as it is slow. This is shown by the fact that the time does not appear in the final expression in equation 2. Instead, Hannay's angle is given by the line integral of a vector $\mathbf{A}$ along the path in parameter space that the system follows. Furthermore, this integral is, in general, path dependent, as may be seen by computing the new vector field $\mathbf{B} = \nabla \times \mathbf{A}$, which in general does not vanish. Indeed, by Stokes' theorem, the net angle $\Delta\theta$ experienced by the system when it is adiabatically transported around a closed loop in parameter space is given by the integral of $\mathbf{B}$ over the bounded surface. The vector $\mathbf{B}$ plays the role of an "angle flux" in parameter space.

The notation chosen here emphasizes the analogy with the vector fields $\mathbf{A}$ and $\mathbf{B}$ in magnetostatics, and, in the same analogy, Hannay's vector fields possess a kind of gauge invariance. The gauge degree of freedom corresponds to a redefinition of the origin of the angle θ in phase space.

THE CASE OF GUIDING CENTER MOTION

In the nonrelativistic motion of a charged particle in a uniform magnetic field, the particle moves in a helical orbit about a magnetic field line with angular frequency $\Omega = eB/mc$, which is called the "gyrofrequency." The average particle position is called the "guiding center," and it moves along the field line with constant velocity. The angle associated with the gyrofrequency is called the "gyrophase," and it may be considered to be the angle between the perpendicular component of the velocity vector and some arbitrary but fixed direction in the perpendicular plane.

Motion in a nonuniform magnetic field, however, is more complicated.[3] Not only does the guiding center gradually drift in a direction perpendicular to the field lines, but also the dynamics of the gyrophase becomes more interesting. An obvious modification to the gyrophase is that in a nonuniform field, the gyrofrequency becomes a function of position, so part of the time evolution of the gyrophase is given by

$$\theta(t) = \int_0^t \Omega(\mathbf{x}(t')) \, dt', \tag{4}$$

where $\mathbf{x}$ is the guiding center position. But there is another contribution $\Delta\theta$ to the gyrophase, which is analogous to Hannay's angle, and which is tied up with the fact that in a nonuniform field it is not possible to choose a constant reference direction in the perpendicular plane to which the definition of the gyrophase should be tied.

In a nonuniform field, it is best to introduce a field of orthonormal triads, $(\hat{e}_1, \hat{e}_2, \hat{b})$, where $\hat{b} = \mathbf{B}/B$, and where $(\hat{e}_1, \hat{e}_2)$ are unit vectors perpendicular to $\hat{b}$. These latter two vectors are not unique, but are subject to a redefinition by a rotation in the perpendicular plane. It is logical to define the gyrophase as the angle between one of the unit vectors and the perpendicular velocity of the particle; if we do so, then the gyrophase is not unique either, but becomes redefined when $(\hat{e}_1, \hat{e}_2)$ are redefined. This redefinition is a kind of a gauge transformation, and is related to the gauge structure found by Hannay.

This gauge structure is revealed when the equation of evolution for the gyrophase is worked out by perturbation theory. The result is

$$\dot{\theta} = \Omega(\mathbf{x}) + \mathbf{R} \cdot \dot{\mathbf{X}} + (\cdots), \qquad (5)$$

where the ellipsis indicates terms of no significance to us, and where $\mathbf{R}$ is defined by

$$\mathbf{R} = \nabla \hat{e}_1 \cdot \hat{e}_2. \qquad (6)$$

The second term of equation 5 corresponds to Hannay's angle, and the vector $\mathbf{R}$ corresponds to Hannay's vector $\mathbf{A}$, for if we integrate this second term, we get

$$\Delta\theta = \int_{\mathbf{x}_0}^{\mathbf{x}_1} \mathbf{R} \cdot d\mathbf{x}, \qquad (7)$$

which shows that $\Delta\theta$ is given by a generally path-dependent line integral, just as in Hannay's analysis.

However, there is more to the phase anholonomy in guiding center motion than this, because the gyrophase is not only an angle in phase space but also in real space. The latter space has a metrical structure, and leads to a transport equation for frames that is much like Fermi–Walker transport in special relativity. We can derive this structure by considering what happens to the vectors $(\hat{e}_1, \hat{e}_2)$ as the guiding center moves from one place to another. Consider, for example, the change in $\hat{e}_1$ as the guiding center moves from $\mathbf{x}$ to $\mathbf{x} + d\mathbf{x}$. This change $d\hat{e}_1$ can have no component along $\hat{e}_1$ itself, because $\hat{e}_1$ is a unit vector. The other two components are given by

$$\begin{aligned}d\hat{e}_1 &= d\mathbf{x} \cdot \nabla \hat{e}_1 \\ &= (d\mathbf{x} \cdot \nabla \hat{e}_1 \cdot \hat{e}_2)\hat{e}_2 + (d\mathbf{x} \cdot \nabla \hat{e}_1 \cdot \hat{b})\hat{b} \\ &= (\mathbf{R} \cdot d\mathbf{x})\hat{e}_2 - (d\mathbf{x} \cdot \nabla \hat{b} \cdot \hat{e}_1)\hat{b}. \end{aligned} \qquad (8)$$

The first term of the last line shows that the quantity $\mathbf{R} \cdot d\mathbf{x}$ represents the infinitesimal angle in the moving perpendicular plane through which $\hat{e}_1$ has rotated, and the second term shows the increment in $\hat{e}_1$ that it must have to remain perpendicular to $\hat{b}$. Therefore, we can say that Hannay's angle in the guiding center problem is the accumulated angle of rotation of $\hat{e}_1$ in the perpendicular plane as the guiding center moves on its trajectory.

One might say that the first term, representing Hannay's angle, occurs in the guiding center problem only because the vectors $(\hat{e}_1, \hat{e}_2)$ were poorly defined. Therefore, let us consider a redefinition of these vectors that will make this first term vanish.

Now instead of filling up a region of space with a field of triads, we will try to make a definition that is as rotationless as possible. We begin by choosing a single triad (not a field) at a single point $\mathbf{x}_0$, and then by extending this definition to points that lie on a curve $\mathbf{x} = \mathbf{x}(s)$ passing through $\mathbf{x}_0$. (Here, s is the distance along the curve.) The extension is accomplished by solving the transport equation.

$$\frac{d\mathbf{u}}{ds} = -\hat{\mathbf{b}} \left(\frac{d\hat{\mathbf{b}}}{ds} \cdot \mathbf{u} \right), \tag{9}$$

where $\mathbf{u}$ is any perpendicular vector, and where the equation comes from adapting the final term of equation 8. This transport equation will guarantee that the triad ($\hat{\mathbf{e}}_1, \hat{\mathbf{e}}_2, \hat{\mathbf{b}}$) will be rotationless as seen by a guiding center moving along the given curve.

This transport equation is a 3-dimensional analog of Fermi–Walker transport, and it guarantees that the vectors transported will indeed remain orthonormal. It also has an interesting connection with parallel transport as it occurs in non-Euclidean geometry. By using this equation to transport a vector around a small parallelogram, one can derive the following expression for the vector $\mathbf{N} = \nabla \times \mathbf{R}$:

$$\mathbf{N} = \tfrac{1}{2} \hat{\mathbf{b}}[(b_{i,j} b_{j,i}) - (\nabla \cdot \hat{\mathbf{b}})^2] + (\nabla \cdot \hat{\mathbf{b}}) \hat{\mathbf{b}} \cdot \nabla \hat{\mathbf{b}} - \hat{\mathbf{b}} \cdot \nabla \hat{\mathbf{b}} \cdot \nabla \hat{\mathbf{b}}. \tag{10}$$

This vector is the analog of Hannay's vector $\mathbf{B}$, and it represents the curvature form of the gauge structure associated with guiding center motion. It has a number of interesting properties that I have discussed elsewhere.[4]

CONCLUSIONS

For applications to guiding center problems and self-consistent theories in plasma physics, the main lesson of this analysis is that the definition of the gyrophase is tied up with a gauge degree of freedom of a novel kind, and that in practical applications it is best to understand this gauge dependence and to use it. I have given a much more careful and complete accounting of the geometrical picture associated with this gauge dependence in another publication.[4] An interesting aspect of this theory is that it bears a remarkably strong similarity to the rotations that occur in Thomas precession. I have also discussed this elsewhere. Indeed, since the discovery of Berry's phase, all the evidence is that gauge theories are not exotic or exceptional in physics, but, with a little attention, they are likely to appear everywhere. The guiding center problem is an example of this.

REFERENCES

1. BERRY, M. V. 1984. Proc. R. Soc. Ser. A **392**: 45.
2. HANNAY, J. H. 1985. J. Phys. A **18**: 221.
3. NORTHROP, T. G. 1963. The Adiabatic Motion of Charged Particles. Interscience. New York.
4. LITTLEJOHN, R. G. 1987. Submitted for publication in Phys. Rev. A.

Integrability of Nonlinear Wave Equations[a]

H. H. CHEN[b] AND J. E. LIN[c,d]

[b]*Laboratory for Plasma and Fusion Energy Studies*
University of Maryland
College Park, Maryland 20742

[c]*Department of Mathematical Sciences*
George Mason University
Fairfax, Virginia 22030

INTRODUCTION

It is now well known that nonlinear evolution equations like the Korteweg–de Vries equation (KdV),

$$u_t = 6uu_x + u_{xxx}, \qquad (1)$$

is completely integrable.[1-8] Gardner et al.[1] first showed that equation 1 is the compatibility condition of two linear equations

$$L\phi = (\partial_x^2 + u)\phi = \lambda\phi \qquad (2a)$$

and

$$A\phi = (4\partial_x^3 + 6u\partial_x + 3u_x)\phi = \phi_t. \qquad (2b)$$

The so-called Lax pair,[2] L and A, satisfies the operator equation

$$L_t = [A, L] = AL - LA. \qquad (3)$$

Historically, equation 2a was found through the Miura[9] transformation,

$$u = M(v) = -(1/2)v_x - (1/4)v^2 + \lambda, \qquad (4)$$

which relates the KdV equation 1 to the modified KdV (MKdV) equation,

$$v_t = -(3/2)v^2 v_x + v_{xxx} + 6\lambda v_x, \qquad (5)$$

and equation 2a was obtained from equation 4 by letting

$$v = 2\phi_x/\phi. \qquad (6)$$

Actually, Miura[9] found this transformation by carefully examining the first few conservation laws that were known for both KdV and MKdV equations. This is

---

[a]The work of one of the authors (H.H.C.) was supported in part by the National Science Foundation.
[d]To whom correspondence should be sent.

certainly not the desired method to search the Lax pair for other integrable equations.

About ten years ago, one of the authors (HHC) together with Y. C. Lee and C. S. Liu[10] had proposed a method to systematically construct the infinite set of conservation laws directly from the linear variational equation (see Definition 3 below) of the original equation. They also used a linear mode-coupling scheme to construct the spectral problem that together with the linear variational equation constitutes the Lax pair for the inverse scattering method. In this paper, we would like to propose a second method to find the spectral problem. The new approach is able to construct the Miura transformation directly and utilize a factorization scheme to construct the linear recursion operator of the symmetries (see Definition 3 below), which is also the spectral operator that we are looking for. The adjoint of this pair consists of the eigenvalue problem of the linear recursion operator of gradients (see Definition 1 below) of constants of motion together with the adjoint of the linear variational equation.

Before we start, let us clarify some of the terminology that are used heavily in this paper.

Definition 1:[2,11] Given a functional $I = \int_{-\infty}^{\infty} q(u)\, dx$, where $u = u(x, \alpha)$ is its argument function that depends on a parameter α as well as on the variable x and vanishes sufficiently fast at the infinity, the *gradient* (functional derivative) of I with respect to u, $\delta I/\delta u$, is defined by

$$(d/d\alpha)\, I\{u\} = \int_{-\infty}^{\infty} (\delta I/\delta u)\, (\partial u/\partial \alpha)\, dx.$$

Definition 2: Given a function $F(u) = F(x, t, u, u_x, \ldots)$ that depends on u and its partial derivatives and possibly on the variables x and t,

$$F'[v] \equiv \partial F(u + \epsilon v)/\partial \epsilon \big|_{\epsilon=0}$$

is the Gateaux derivative of F in the direction v with respect to u. Therefore, the operator F' that depends on u acts on the function v.

Definition 3: Given a nonlinear evolution equation,

$$u_t = K(u),$$

where $K(u)$ depends on u and its partial derivatives and possibly on the variables x and t, a function $\sigma(x, t, u, u_x, \ldots)$ is called a *symmetry* of the equation if σ satisfies the *linear variational equation* of $u_t = K(u)$:

$$(\partial/\partial t)\sigma = K'[\sigma], \qquad (7)$$

where $K'[\sigma]$ is defined as in Definition 2. These symmetries are the infinitesimal generators of one-parameter groups of invariant transformations of the equation. A classical theorem of Noether[12] states that constants of motion can usually be constructed from the symmetries of the Lagrangian.

We will denote the adjoint of an operator N by N^*. All the solutions of equations in the following work are assumed to vanish sufficiently fast at infinity unless otherwise stated.

THE NEW APPROACH

Background

Given an evolution equation $u_t = K(u)$, it is known[2] that the gradient ψ of the constants of motion of this equation satisfies the adjoint of the linear variational equation 7 of the original equation:

$$\psi_t = -K'^*(\psi). \tag{8}$$

On the other hand, it is known that an integrable equation possesses an infinite set of conservation laws; therefore, there must be infinitely many polynomial solutions $\psi = \psi$ $(x, t, u, u_x, \ldots)$ of equation 8. Taking equation 8 as one member of the Lax equations, a method to construct infinitely many constants of motion for the integrable nonlinear evolution equation is proposed in reference 10. They used a WKB-type expansion of the solution ψ of equation 8,

$$\psi = \exp\left(kx + \omega t + \int_{-\infty}^{x} T\, dx\right) \tag{9}$$

with $T = T(u) = \sum_{n=0}^{\infty} k^{-n} T_n$, where k and ω satisfy the linear dispersion relation. Take the KdV equation 1 as an example—equation 8 becomes

$$\psi_t = 6u\psi_x + \psi_{xxx}. \tag{10}$$

Substituting equation 9 into equation 10 yields a recursive relation for T_n:

$$\int_{-\infty}^{x} (T_n)_t\, dx = 6u\delta_{n+1,0} + 6uT_n + (T_n)_{xx} + 3(T_{n+1})_x + 3\sum_{j=0}^{n} T_j(T_{n-j})_x$$
$$+ \sum_{j+m=0}^{n} T_j T_m T_{n-j-m} + 3\sum_{j=0}^{n+1} T_j T_{n+1-j} + 3T_{n+2}, \tag{11}$$

where we use $\delta_{i,j}$ to denote the Kronecker delta. The preceding formula can be used to generate T_{n+2} (the last term on the right-hand side) from the knowledge of $T_j, j = 0, 1, 2, \ldots, n + 1$. Assuming $T_{-2} = 0 = T_{-1}$, we have, from equation 11,

$T_0 = 0$,

$T_1 = -2u$,

$T_2 = 2u_x$,

$T_3 = -2u^2 - 2u_{xx}$,

$T_4 = (4u^2 + 2u_{xx})_x$,

$T_5 = -2u_{xxxx} - 10u_x^2 - 12uu_{xx} - 4u^3$,

etc. Note that if $T_0, T_1, \ldots, T_{n+2}$ are differential polynomials of u, that $\int_{-\infty}^{\infty} (T_n)_t\, dx = 0$, provided u vanishes sufficiently fast at $x = \pm\infty$ and we have a conservation law $\int_{-\infty}^{\infty} T_n\, dx$ = constant. If this process can go on forever, we then have an infinite set of conservation laws, and whether this process can go on depends on whether $\int_{-\infty}^{x} T_n \, dx$ can be expressed in terms of differential polynomials of u. In fact, for general nonlinear evolution equations, this criterion serves as a test to distinguish an integrable system from a nonintegrable one. For example, the quartic KdV equation

$$w_t = 6w^3 w_x + w_{xxx}$$

would not pass this test. We can show immediately that

$$T_1 \text{ (quartic KdV equation)} = -2w^3,$$

but w^3 is not a conserved density at all. The recursive formula therefore is disrupted at $n = 1$. On the other hand, for the integrable system like the KdV equation 1, we can go on to find the higher members of the conserved densities without disruption. However, from equation 11 alone, we cannot establish the fact that the recursive operation could go on forever. More information is needed to do that. If we can find a nontrivial linear recursion operator of the solutions of equation 10, which are gradients of constants of motion for the KdV equation 1, then this would be enough to establish the existence of infinitely many conservation laws, provided there is at least one conservation law for the equation. It is therefore our goal in this paper to construct this linear recursion operator, which is also the spectral operator[10] for a pair of Lax operators.

A New Twist

We would like to extend equation 9, reinterpret it, and construct from it a linear recursion operator of the gradients of constants of motion for the KdV equation 1. This operator can be taken as the spectral operator L in the Lax pair, equation 3, with a proper operator A.

Note first that equation 9 implies that $\psi_x - p\psi = 0$, where $p = k + T$. We now propose to add a new term ξ to the preceding equation for ψ,

$$\psi_x - p\psi = \xi. \tag{12}$$

Proposition: *ξ is a solution of the adjoint of the linear variational equation of another nonlinear evolution equation.*

Demanding that equation 12 be compatible with equation 10, we obtain

$$p_t = 6u_x p + 6u p_x + p_{xxx} + 3p_{xx} p + 3p^2 p_x + 3p_x^2 \tag{13}$$

and

$$\xi_t = \xi_{xxx} + (6u + 3p_x)\xi_x + (6u_x + 3p_{xx} + 3p_x p)\xi. \tag{14}$$

We now assume that equation 14 is the adjoint of the linear variational equation of a nonlinear evolution equation, say,

$$v_t = m(v, v_x, \ldots). \tag{15}$$

Of course, $u = u(v, v_x, \ldots)$ and $p = p(v, v_x, \ldots)$. Taking the adjoint of equation 14, we get

$$\eta_t = \eta_{xxx} + (6u + 3p_x)\eta_x - 3p_x p\eta$$

which should be the linear variational equation of equation 15. Thus we have

$$v_t = v_{xxx} + F(v, v_x)$$

for some function F. We immediately see that

$$\begin{cases} \partial F/\partial v = -3pp_x & \text{(16a)} \\ \partial F/\partial v_x = 6u + 3p_x. & \text{(16b)} \end{cases}$$

It follows that $u = u(v, v_x)$ depends only on v and v_x and that $p = p(v)$ depends only on v. The compatibility of equations 16a and 16b now yields

$$-3pp' = 6(\partial u/\partial v) + 3p''v_{x'}$$

where $p' = dp/dv$, $p'' = d^2p/dv^2$, and $p^{(n)} = d^np/dv^n$. Hence

$$u = -p^2/4 - p_x/2 + f(v_x) \tag{17}$$

for some function f. Also from equation 16a, we have

$$F = -(3/2)p^2 v_x + g(v_x)$$

for some function g. From equations 16a and 17, we have

$$(3/2)p^2 + dg/dv_x = 6u + 3p_x = -(3/2)p^2 + 6f(v_x).$$

Hence

$$dg/dv_x = 6f(v_x). \tag{18}$$

Recall that the equation for p, given in equation 13,

$$p_t = 6u_x p + 6up_x + p_{xxx} + 3p_{xx}p + 3p_x^2 + 3p^2 p_x$$
$$= -(3/2)p^2 p_x + p_{xxx} + 6(pf)_x, \tag{19}$$

has to be consistent with

$$v_t - v_{xxx} = F(v, v_x) = -(3/2)p^2 v_x + g(v_x). \tag{20}$$

Substituting $p = p(v)$ into equation 19, we get

$$v_t - v_{xxx} = -(3/2)p^2 v_x + (p'''/p') v_x^3 + 6v_x f$$
$$+ 3((p''/p')v_x + 2(p/p')(df/dv_x))v_{xx}. \tag{21}$$

Hence, we have from equations 20 and 21

$$g(v_x) = (p'''/p')v_x^3 + 6v_x f \tag{22a}$$
$$p''v_x + 2(df/dv_x)p = 0. \tag{22b}$$

Now, equations 18 and 22a imply that

$$3(p'''/p')v_x^2 + 6v_x(df/dv_x) = 0;$$

hence, from equation 22b,

$$p''p' = pp'''. \tag{23}$$

On the other hand, equation 22b implies

$$p''/p = -(2/v_x)(df/dv_x) = c, \tag{24}$$

where c is a constant, since the left-hand side and the middle term of equation 24 depend only on v and v_x, respectively. Note that equation 24 implies equation 23. Thus we have

$$f = -(c/4)v_x^2 + \lambda \text{ for some constant } \lambda.$$

Equation 17 and equation 21 then imply that

$$u = -(p^2/4) - (p_x/2) - (c/4)v_x^2 + \lambda \tag{25}$$

and

$$v_t - v_{xxx} = -(3/2)p^2 v_x - (c/2)v_x^3 + 6\lambda v_x, \tag{26}$$

respectively. In order for equation 25 to be a Miura transformation relating equation 1 to equation 26, that is to say,

$$u_t - 6uu_x - u_{xxx} = M'(v_t - v_{xxx} + (3/2)p^2 v_x + (c/2)v_x^3 - 6\lambda v_x),$$

where $M' = -(1/2)pp' - (1/2)p''v_x - (1/2)p'\partial_x - (c/2)v_x\partial_x$, we need $c = 0$. Hence, we have $p'' = 0$, that is to say, $p = av + b$ for some constants a and b. Therefore, equation 25 becomes the Miura transformation equation 4,

$$u = -(1/4)(av + b)^2 - (1/2)av_x + \lambda,$$

and equation 26 becomes the MKdV equation 5,

$$v_t - v_{xxx} = -(3/2)(av + b)^2 v_x + 6\lambda v_x.$$

A CONSTRUCTION OF LAX PAIR OPERATORS

Lemma 1: Suppose that L, A, and B are operators depending on t and related by $L_t = BL - LA$. Then $L\eta$ satisfies $\xi_t = B\xi$ if η satisfies $\eta_t = A\eta$.

Proof: $(L\eta)_t = L_t\eta + L\eta_t = (BL - LA)\eta + LA\eta = BL\eta$.

Lemma 2: Suppose that two evolution equations $u_t = K(u)$ and $v_t = G(v)$ are related by a transformation $u = M(v)$, then (a) M' maps the symmetries of $v_t = G(v)$ to those of $u_t = K(u)$ and (b) M'^* maps the gradients of constants of motion of $u_t = K(u)$ to those of $v_t = G(v)$.

Proof: (a) Let $u \to u + \epsilon\sigma$ and $v \to v + \epsilon\theta$ be two infinitesimal transformations of the solutions of $u_t = K(u)$ and $v_t = G(v)$, respectively. Then, from $u = M(v)$, we obtain

$$\sigma = M'\theta. \tag{27}$$

Note that σ and θ are symmetries (Definition 3) of $u_t = K(u)$ and $v_t = G(v)$, respectively. Thus we have obtained (a).

(b) Differentiating with respect to t on both sides of equation 27, we get

$$\sigma_t = (M'\theta)_t = (M')_t\theta + M'\theta_t.$$

Now, by the definition of symmetries, σ and θ satisfy $\sigma_t = K'\sigma$ and $\theta_t = G'\theta$, respectively. Hence

$$K'\sigma = (M')_t\theta + M'G'\theta.$$

that is to say,

$$K'M' = (M')_t + M'G'.$$

Taking adjoints on both sides, we obtain

$$(M'^*)_t = M'^*K'^* - G'^*M'^* = (-G'^*)M'^* - M'^*(-K'^*).$$

By Lemma 1, $M'^*\eta$ satisfies $\xi_t = -G'^*\xi$, if η satisfies $\eta_t = -K'^*\eta$. By a result of Lax,[2] $\eta = \eta(x, t, u, u_x, \ldots)$ and $\xi = \xi(x, t, v, v_x, \ldots)$ are the gradients of constants of motion of $u_t = K(u)$ and $v_t = G(v)$, respectively. Thus (b) is proved.

Theorem. In addition to the assumption of Lemma 2, if equations $u_t = K(u)$ and $v_t = G(v)$ have independent Hamiltonian structures (i.e., none of them can be derived from the other by the relation $u = M(v)$), then we can construct the linear recursion operators of symmetries and gradients of constants of motion for both equations. For example, if we let J_1 and J_2 be the Hamiltonian operators of these two Hamiltonian systems for $u_t = K(u)$ and $v_t = G(v)$, respectively, then

$$R_s = M'J_2M'^*J_1^{-1} \quad \text{and} \quad U_s = J_2M'^*J_1^{-1}M'$$

are linear recursion operators of symmetries of equations $u_t = K(u)$ and $v_t = G(v)$, respectively, and

$$R_a = J_1^{-1}M'J_2M'^* \quad \text{and} \quad U_a = M'^*J_1^{-1}M'J_2$$

are linear recursion operators of gradients of constants of motion of equations $u_t = K(u)$ and $v_t = G(v)$, respectively.

Proof: Since the Hamiltonian operator maps the gradients of constants of motion of the equation to its symmetries, we see from Lemma 2 that symmetries of $u_t = K(u)$ are mapped by J_1^{-1} to the gradients of constants of motion of $u_t = K(u)$, which are then mapped by M'^* to the gradients of constants of motion of $v_t = G(v)$, which are then mapped by J_2 to the symmetries of $v_t = G(v)$, which are then mapped by M' to the symmetries of $u_t = K(u)$. Thus, $R_s = M'J_2M'^*J_1^{-1}$ maps symmetries of $u_t = K(u)$ to those of $u_t = K(u)$; that is to say, R_s, which is obviously linear, is a recursion operator of symmetries for $u_t = K(u)$. This proves part of the assertion. The rest of the assertion can be proved similarly.

Remark 1: By the work of Chen et al.,[10] we have the following pairs of *Lax operators* 3 for $u_t = K(u)$: R_s with K' and R_a with $-K'^*$; that is to say,

$$(\partial/\partial t) R_s = [K', R_s] = K'R_s - R_s K'$$

and

$$(\partial/\partial t) R_a = [-K'^*, R_a] = -K'^* R_a - R_a(-K'^*).$$

Likewise, U_s with G' and U_a with $-G'^*$ are two pairs of Lax operators for $v_t = G(v)$.

Now, the KdV and MKdV equations 1 and 5 are related by the Miura transformation 4 and, moreover, they have the Hamiltonian structures $u_t = J_1 (\delta H_1/\delta u)$ and $v_t = J_2(\delta H_2/\delta v)$, respectively, where the Hamiltonian operators

$$J_1 = \partial_x = J_2$$

and the Hamiltonians

$$H_1 = \int_{-\infty}^{\infty} (u^3 - (\tfrac{1}{2})u_x^2) \, dx \quad \text{and} \quad H_2 = \int_{-\infty}^{\infty} (-(\tfrac{1}{8})v^4 - (\tfrac{1}{2})v_x^2 + 3\lambda v^2) \, dx.$$

Therefore, from the result of the preceding theorem, we have the following three corollaries.

Corollary 1: $R_s = M'J_2 M'^* J_1^{-1} = -(\tfrac{1}{4})(\partial_x^2 + 2u_x \partial_x^{-1} + 4u) + \lambda$, where $M' = -(\tfrac{1}{2})(\partial_x + v)$ and is derived from equation 4, is a linear recursion operator of symmetries of the KdV equation 1.

Corollary 2: $R_a = J_1^{-1} M' J_2 M'^* = -(\tfrac{1}{4})(\partial_x^2 - 2\partial_x^{-1} \cdot u_x + 4u) + \lambda$ is a recursion operator of gradients of constants of motion of the KdV equation 1.

Corollary 3: Similarly, $U_s = J_2 M'^* J_1^{-1} M' = -(\tfrac{1}{4})(\partial_x^2 - v_x \partial_x^{-1} \cdot v - v^2)$ and $U_a = M'^* J_1^{-1} M' J_2 = -(\tfrac{1}{4})(\partial_x^2 + v \partial_x^{-1} \cdot v_x - v^2)$ are linear recursion operators of symmetries and gradients of constants of motion of the MKdV equation 5, respectively.

Remark 2: From Remark 1, we now have the following two pairs of Lax operators for the KdV equation 1:

(i) R_s with $A = 6u_x + 6u\partial_x + \partial_x^3$ and
(ii) R_a with $B = -A^* = 6u\partial_x + \partial_x^3$.

Likewise, we also have the following two pairs of Lax operators for the MKdV equation 5:

(iii) U_s with $A_m = -3vv_x - (\tfrac{3}{2})v^2 \partial_x + \partial_x^3 + 6\lambda \partial_x$ and
(iv) U_a with $B_m = -A_m^* = -(\tfrac{3}{2})v^2 \partial_x + \partial_x^3 + 6\lambda \partial_x$.

Remark 3: The set of the Lax scattering equations 2a and 2b associated with the Lax operators in Remark 2 can be reduced to the more familiar set of the Lax scattering equations.[1-8] For example, the set of the Lax scattering equations associated with (i) of Remark 2,

$$\begin{cases} \phi_{xx} + 2u_x \partial_x^{-1} \phi + 4u\phi = 4\lambda \phi \\ \phi_t = \phi_{xxx} + 6u\phi_x + 6u_x \phi, \end{cases}$$

is related to the more familiar set,[1,2]

$$\begin{cases} \psi_{xx} + u\psi = \lambda\psi \\ \psi_t = 4\psi_{xxx} + 6u\psi_x + 3u_x\psi, \end{cases}$$

by

$$\phi = \psi\psi_x.$$

SUMMARY

In summary, we have demonstrated a new method to construct the spectral problem of the Lax scattering equations for solving an integrable nonlinear evolution equation such as the KdV equation. The application of this method to the other integrable nonlinear evolution equations will be published elsewhere.

REFERENCES

1. GARDNER, C., J. GREENE, M. KRUSKAL & R. MIURA. 1974. Commun. Pure Appl. Math. **27**: 97–133.
2. LAX, P. 1968. Commun. Pure Appl. Math. **21**: 467–490.
3. SCOTT, A. C., F. Y. F. CHU & D. W. MCLAUGHLIN. 1973. Proc. IEEE **61**: 1443–1483.
4. ABLOWITZ, M. J., D. J. KAUP, A. C. NEWELL & H. SEGUR. 1974. Studies Appl. Math. **53**: 249–315.
5. MIURA, R. M. 1976. SIAM Rev. **18**: 412–459.
6. ABLOWITZ, M. J. & H. SEGUR. 1981. Solitons and the Inverse Scattering Transform. Society for Industrial and Applied Mathematics. Philadelphia.
7. NEWELL, A. C. 1985. Solitons in Mathematics and Physics. Society for Industrial and Applied Mathematics. Philadelphia.
8. NOVIKOV, S., S. MANAKOV, L. PITAEVSKII & V. E. ZAKHAROV. 1984. Theory of Solitons. Consultants Bureau. New York.
9. MIURA, R. M. 1968. J. Math. Phys. **9**: 1202–1204.
10. CHEN, H. H., Y. C. LEE & C. S. LIU. 1979. Phys. Scr. **20**: 490–492.
11. GARDNER, C. S. 1971. J. Math. Phys. **12**: 1548–1551.
12. ARNOLD, V. I. 1978. Mathematical Methods of Classical Mechanics. Springer-Verlag. New York/Berlin.

Normalization in the Face of Integrability

ANDRÉ DEPRIT[a] AND BRUCE R. MILLER

National Bureau of Standards
Gaithersburg, Maryland 20899

> Magic casements, opening on the foam
> Of perilous seas, in faery lands forlorn.
> —JOHN KEATS

THE ILLUSION OF INTEGRABILITY

Celestial mechanics is a pleasant occupation, although at times fraught with real hazards. We are peddlers of illusions; chief among them is the illusion of integrability. That most perturbed Keplerian systems are not integrable we are not loath to admit nor are we embarrassed to confess that we do not know any other way of studying them than by making them integrable. Perturbations break symmetries; it is their nature, no matter how small they are. There are exceptions, of course, but they are rare indeed. Unavoidably a violation of symmetry results in the loss of integrability. Nonetheless, we like to convince ourselves that the disturbance is small enough that we can replace the original system with an approximation to which we restore some of the lost symmetries, enough of them to make the substitute integrable.

The fakery goes nowadays by the respectable name of *normalization*. It came into the light of day as an averaging process. Concerned as they were with the task of fitting sparse, widely spread observations to a nominal trajectory, astronomers learned to regard the perturbations as having two kinds of effects, the short period fluctuations modulating the long trend variations, a reenacting if you will of Ptolemy's fiction of fast epicycles rolling over roulettes in slow drift. By marching in large steps along a smoothed path, the integration of the averaged equations connects far apart observations. A high moment in the development of mechanics was reached when Delaunay equated the method of averaging to a canonical transformation that renders the mean anomaly l ignorable in the Hamiltonian representing a perturbed Keplerian system. The angle l being eliminated, its conjugate momentum L becomes an integral. But integrals beget symmetries. With enough angles eliminated from the original system, enough symmetries will have been imposed on the transformed Hamiltonian to make it integrable.

Our working conditions have changed radically: telemetry data pouring down from satellites in dense showers ensure an almost uniform covering of their paths. Our interest has shifted nowadays to the task of designing trajectories to satisfy mission requirements. Yet the problem of separating short-term effects from long-trend evolutions is still with us in so far as we try to tell aerospace designers how close their satellites will adhere for how long to the anticipated track.

---

[a]To whom correspondence should be addressed.

We invent trajectories, we are therefore interested in knowing the totality of possible trajectories. This global view we obtain by a process called *reduction*. Dynamical symmetries are the keystones of reductions. They afford a partition of the universe of all possible solutions into classes of equivalence, two orbits being equivalent if one is the reproduction of the other by a symmetry. In the old days, equivalence classes of orbits in perturbed Keplerian problems went by the strange name of *osculating ellipses*. Strange name indeed considering that these are not simply ellipses nor are they osculating curves by the definitions of geometry. At the root of the concept is the notion that a quintuple (L, G, H, g, h) of Delaunay variables represents a curve made of an ellipse in the coordinate space and of a circle—Hamilton's hodograph—in the space of velocities, hence that the mapping $\phi : l \mapsto l + \epsilon$ acting like a rotation in the 6-dimensional phase space applies this phase curve upon itself. For a pure Keplerian system, ϕ is a symmetry. But, in the overwhelming majority of perturbed Keplerian systems, that symmetry is broken because, as classical astronomers are prompt to observe, the perturbation depends usually on the mean anomaly. In which case we play, if we feel desperate enough, the magic trick of replacing the original system with one that we hope is almost like it, but that enjoys the Delaunay rotations in mean anomaly as its group of symmetries.

Under well-specified conditions to be satisfied by the group of symmetries, the equivalence classes may be handled like points in the phase space of the normalized Hamiltonian. Classical mechanics teaches how to reach the orbital phase space at least in an extrinsic manner via a system of coordinates and conjugate momenta, one of the coordinates to be ignorable in the reduced Hamiltonian. Past the formal procedure of removing a degree of freedom by ignoring a coordinate, not until recently had classical mechanics paid much attention to the intrinsic (geometric) properties of the reduced phase space. The situation has improved considerably in the last few years. For instance, in the case of a perturbed elliptic oscillator (two coupled harmonic oscillators of equal fundamental frequencies), it has been recognized that each manifold $L =$ constant in the reduced phase space is a sphere, hence that the orbits of the reduced system are the level contours on that sphere of the Hamiltonian representing the reduced system.

The global behavior of the system is commanded to a large extent by its equilibria. Of particular interest are the unstable equilibria because of the manifolds of asymptotic orbits emanating from them. It is very hard to decide numerically whether unstable equilibria are linked by homoclinic asymptotic orbits. As a matter of fact, it is generally presumed that this happens only very exceptionally, usually in reduced systems that are too simple to be realistic. But, if the manifold of asymptotic orbits going out of an unstable equilibrium crosses the manifold of asymptotic orbits coming to an unstable equilibrium, then all sorts of extraordinary events are expected to happen along the heteroclinic solutions. An unstable equilibrium stands like a landmark above the epicenter of a zone of chaotic dislocations. In that respect, unstable equilibria are symptoms that the normalized system that spawns them may not be a faithful photograph of the unreduced system.

Overall, normalization acts like a polarizing filter: the panorama is blurred in spots by haze and mist, but the symmetrizing lens discounts the fog, although it is as much a part of the reality as the landscape it veils. About photographers, illusionists, and dynamical astronomers, the question seems unavoidable: Who is fooling whom?

A TEST BENCH

A dynamical system capsulizes the debate better than most. It is the three-particle Toda lattice.[1,2] With

$$\mathcal{V} = e^{p_1-p_2} + e^{p_2-p_3} + e^{p_3-p_1} \tag{1}$$

as the potential energy, it is described by the Hamiltonian

$$\mathcal{H} = \tfrac{1}{2}(P_1^2 + P_2^2 + P_3^2) + \mathcal{V}. \tag{2}$$

Recent analyses of the three-particle Toda lattice point to three key features:

(a) The system admits three independent integrals in involution, for instance, the functions

$$\mathcal{J}_1 = P_1 + P_2 + P_3, \tag{3}$$

$$\mathcal{J}_2 = P_1 P_2 + P_2 P_3 + P_3 P_1 - \mathcal{V}, \tag{4}$$

$$\mathcal{J}_3 = P_1 P_2 P_3 - P_1 e^{p_2-p_3} - P_2 e^{p_3-p_1} - P_3 e^{p_1-p_2}. \tag{5}$$

These integrals are in involution, and they are independent. Hence the system is integrable.[3,4]

Incidentally, the energy integral $\mathcal{H}$ is the combination $\mathcal{H} = \tfrac{1}{2}\mathcal{J}_1^2 - \mathcal{J}_2$ of the basic integrals.

(b) A reduction (in the technical sense of the word) based on the integral $\mathcal{J}_1$ converts the three-particle Toda lattice into a Hamiltonian system with only two degrees of freedom.[5] The reduced system admits a second integral, and is therefore integrable.

(c) The origin in the phase space of $\mathcal{H}$ is an equilibrium. Let $\mathcal{T}_{n-2}$ denote $\mathcal{H}$ developed as a Taylor series in the neighborhood of the equilibrium, and truncated at degree $n (\geq 2)$, that is, all terms of degree $m > n$ are omitted from the expansion. It has been proved very recently that, for any $n > 2$, the Hamiltonian $\mathcal{T}_n$ is not integrable.[6,7]

In the face of these results, it looks as though the normalization process, because it purports to approximate each $\mathcal{T}_n$ by an integrable Hamiltonian, say $\mathcal{T}'_n$, is an exercise in futility. Not unexpectedly, a sophomoric viewpoint leads to a contrary conclusion. Indeed, as we shall show, much is to be learned on how the dynamical systems defined by the various Hamiltonians $\mathcal{H}$, $\mathcal{T}_n$, and $\mathcal{T}'_n$ are related to one another from examining the sequence of phase spaces emanating from $\mathcal{T}'_n$ as n tends to infinity.

For the sake of clarity, we introduce a small parameter ϵ as an indicator of the order of the approximations. For example, after it has been developed as a Taylor series at the equilibrium, the reduced Hamiltonian $\mathcal{H}$ is set as the power series

$$\mathcal{H} = \sum_{n\geq 0} \epsilon^n \mathcal{H}_n, \tag{6}$$

where $\mathcal{H}_0$ is a quadratic form in the coordinates and momenta whereas, for any $n > 0$, $\mathcal{H}_n$ is a homogeneous polynomial of degree $n + 2$ in the coordinates. We refer to the exponent of ϵ as the *order* of the approximation. To be quite clear, whereas some authors use interchangeably the words *degree* and *order* to mean the same thing, we use the term *order* in reference to the asymptotic scale established by equation 6, which makes us hold homogeneous polynomials of degree n as terms of order $n - 2$.

By our conventions,

$$\mathcal{T}_n = \sum_{0 \leq m \leq n} \epsilon^m \mathcal{H}_m$$

The normalization converts equation 6 into a series

$$\mathcal{N} = \sum_{n \geq 0} \epsilon^{2n} \mathcal{N}_{2n} \tag{7}$$

whose coefficients belong to the kernel of the Lie derivative

$$L_0: F \mapsto (F; \mathcal{N}_0).$$

As a result, $(\mathcal{N}; \mathcal{N}_0) = 0$, hence $\mathcal{N}_0$ is formally an integral of the normalized system. The normalization is built as a Lie transformation,[8] that is to say, as a canonical transformation developed in the powers of the small parameter ϵ. We also normalize each truncated Hamiltonian $\mathcal{T}_n$. A straightforward analysis of the operations involved in constructing each term in the normalized Hamiltonian shows that, for each $n > 0$, the truncated Hamiltonian $\mathcal{T}_n$ is converted into a formal series of the form

$$\mathcal{T}'_n = \sum_{0 \leq 2m \leq n} \epsilon^{2m} \mathcal{N}_{2m} + \sum_{2m > n} \epsilon^{2m} \mathcal{T}'_{n,2m}. \tag{8}$$

The canonical map in which the normalization will be executed will reveal that, on each manifold $\mathcal{N}_0 = $ constant, the phase space of any $\mathcal{T}'_n$ is a two-dimensional sphere. The orbits of the reduced systems in the manifolds $\mathcal{N}_0 = $ constant are the level curves of $\mathcal{T}'_n$ on that sphere. At whatever order the normalization is carried out for whatever degree of truncation, it will be shown that there exist two isolated equilibria on the orbital sphere; both are *stable*. In the phase plane of average coordinates, they correspond to a unique circle, but traveled either in the direct or the retrograde sense. These critical points being located at the opposite ends of a diameter, we make the picture clearer if we refer to this diameter as the axis of the sphere. What we shall designate as the equatorial circle is the intersection of the sphere with the plane through the center perpendicular to the axis of the sphere. Points on the equator stand for classes of orbits which, on the average, are rectilinear.

The equatorial circle is the siege of momentous events because, as we shall learn, it happens to be the only place other than the poles where, for any n, a normalized Hamiltonian $\mathcal{T}'_n$ could admit an equilibrium. How critical points appear and disappear along the equator as n increases will make an interesting story.

If all terms of order higher than n are dropped from the right-hand member of equation 8, the Hamiltonian so truncated represents a *degenerate* dynamical system in the sense that it admits nonisolated equilibria. Indeed, as our developments covering the interval $1 \leq n \leq 18$ will show, any point on the equator is an equilibrium for such a partial normal form. But, for any $n \geq 2$, the next term in the normal form $\mathcal{T}'_n$, that is the coefficient $\mathcal{T}'_{n,2n'}$, where n' is the smallest integer such that $2n' > n$, breaks the degeneracy to leave only six isolated equilibria. Once the disintegration has occurred, terms of higher order in the normalization like $T'_{n,2n'+2}$, $T'_{n,2n'+4}$, etc., cannot reverse the process. For $n = 1$, T_1 is the Hamiltonian of the Hénon-Heiles oscillator. Its second-order normalization, $\mathcal{N}_0 + \epsilon^2 T'_{1,2}$ is still degenerate, but all equilibria on the equator save six points evaporate at the fourth order under the action of $T'_{1,4}$. A similar study conducted by Carles Simó[9]—but only for $1 \leq n \leq 13$ and in the awesome coordinates of Braun[10]—corroborates our conclusions. That degeneracies at a given order of the normalization may be broken by terms of a higher order is a well-known phenomenon in nonlinear dynamics.[11]

In one respect, the three-particle Toda lattice is not a typical dynamical system, or to say it the way mathematicians talk nowadays, it is not a *generic* problem. For it admits a discrete group of symmetries. Reduction to two degrees of freedom by means of the integral $\mathcal{J}_1$ and normalization may be carried out in a manner that preserves these discrete symmetries. The fact that, after the first reduction, the normalized system is invariant for the rotations by 120° in both the plane of coordinates and the plane of momenta explains why, for each truncation T_n, the breakdown of the degeneracy results into six isolated equilibria at the vertices of a regular hexagon. Three of the equilibria are stable, and they are separated from one another by three unstable equilibria.

Because the truncated normalized systems T'_n are integrable, the asymptotic manifolds emanating from the unstable equilibria continue as homoclinic orbits asymptotic to the unstable equilibria when t goes to $\pm\infty$. Since the truncated, but unnormalized, system T_n is not integrable, one would expect that the difference $T_n - T'_n$ acting as perturbation on T'_n destroys this elementary behavior. We have not tried to check that this is indeed the case. Looking for asymptotic solutions connecting two unstable equilibria does not seem to be an easy task. The critical points appear in the equator at orders so high that we must give up obtaining global expressions for the asymptotic solutions emanating from them. Yet many experiences with normalization support the view that the presence of isolated unstable equilibria in the normalization of a dynamical system indicates that the original system is not integrable. These singularities act like *hot points* in the normalized substratum; their presence is felt at the surface of the original problem as zones of stochasticity. Some Poincaré cross sections computed recently[12] for the three-particle Toda lattice unmistakably exhibit incipient chaos above unstable equilibria of the normalized system.

SYMMETRIES

To clear the approaches, we begin by recognizing that the three-particle Toda lattice admits a discrete group of symmetries isomorphic to the group of the triangle. We build that invariance in the canonical transformation intended to reduce by one unit the number of degrees of freedom. Since the transformation is applicable to a wide class of systems admitting the symmetry of the triangle, we go into some details.

Hamiltonian equation 2 does not change under any of the coordinate transformations

$$t_0 : q_1 = p_1, \quad q_2 = p_2, \quad q_3 = p_3;$$

$$t_1 : q_1 = p_2, \quad q_2 = p_3, \quad q_3 = p_1;$$

$$t_2: q_1 = p_3, \quad q_2 = p_1, \quad q_3 = p_2;$$
$$t_3: q_1 = -p_2, \quad q_2 = -p_1, \quad q_3 = -p_3;$$
$$t_4: q_1 = -p_1, \quad q_2 = -p_3, \quad q_3 = -p_2;$$
$$t_5: q_1 = -p_3, \quad q_2 = -p_2, \quad q_3 = -p_1.$$

The first three of these transformations correspond to an even permutation of the subscripts (1, 2, 3), and the last three to the central symmetry $q_1 = -p_1, q_2 = -p_2, q_3 = -p_3$ followed by an odd permutation of the subscripts (1, 2, 3). Each transformation t_i is readily extended into a canonical transformation as the product

$$T_i: (p_1, p_2, p_3, P_1, P_2, P_3) \mapsto (t_i(p_1, p_2, p_3), t_i(P_1, P_2, P_3)).$$

According to the following table:

	T_0	T_1	T_2	T_3	T_4	T_5
T_0	T_0	T_1	T_2	T_3	T_4	T_5
T_1	T_1	T_2	T_0	T_4	T_5	T_3
T_2	T_2	T_0	T_1	T_5	T_3	T_4
T_3	T_3	T_5	T_4	T_0	T_2	T_1
T_4	T_4	T_3	T_5	T_1	T_0	T_2
T_5	T_5	T_4	T_3	T_2	T_1	T_0

the transformation T_i form a group isomorphic to the group $\mathcal{S}_3$ of permutations of the set $\{1, 2, 3\}$.

In addition to the discrete group $\mathcal{S}_3$, the three-particle Toda lattice admits a two-parameter group $\mathcal{L}$ of dynamical symmetries consisting of the linear transformations $\phi_{a,A}$ defined by the equations:

$$\phi_{a,A}: P_i = Q_i + A, \quad p_i = q_i + At + a, \quad 1 \leq i \leq 3. \tag{9}$$

For any pair (a, A), $\phi_{a,A}$ is symplectic, and it has a multiplier equal to unity. Furthermore, because it depends on the time, it has a nonvanishing remainder $\mathcal{R} = \mathcal{R}(q_1, q_2, q_3, Q_1, Q_2, Q_3)$. The latter is a complete solution of the system of partial differential equations

$$\frac{\partial \mathcal{R}}{\partial q_i} = -\frac{\partial Q_i}{\partial t} = 0, \quad \frac{\partial \mathcal{R}}{\partial Q_i} = \frac{\partial q_i}{\partial t} = -A, \quad (1 \leq i \leq 3).$$

Visibly, up to the addition of an arbitrary function in the parameters a and A, the function

$$\mathcal{R} = -A(Q_1 + Q_2 + Q_3)$$

is such a solution. Accordingly, any $\phi_{a,A}$ changes the Hamiltonian equations derived from equation 2 into the Hamiltonian equations derived from the function

$$\mathcal{H}' = \mathcal{H} + \mathcal{R} = \tfrac{1}{2}(Q_1^2 + Q_2^2 + Q_3^2) + e^{q_1-q_2} + e^{q_2-q_3} + e^{q_3-q_1} + \tfrac{1}{2}A^2.$$

Therefore, $\phi_{a,A}$ is a symmetry of the Toda lattice. The linear transformations $\phi_{a,A}$ form a two-parameter group; we shall denote it as $\mathcal{L}$.

The group $\mathcal{L}$ is related to the integral 3. Indeed every lattice symmetry may be represented as the contact transformation

$$p_i = q_i + (q_i; \mathcal{W}), \qquad P_i = Q_i + (Q_i; \mathcal{W}), \qquad (1 \le i \le 3) \tag{10}$$

generated by the function

$$\mathcal{W} = (tA + a)(Q_1 + Q_2 + Q_3) - A(q_1 + q_2 + q_3). \tag{11}$$

According to Liouville's theorem,

$$\frac{d\mathcal{W}}{dt} = (\mathcal{W}; \mathcal{H}') + \frac{\partial \mathcal{W}}{\partial t}$$

in the phase flow induced by $\mathcal{H}'$. As we compute the terms in that sum, we find that $d\mathcal{W}/dt = 0$, or that $\mathcal{W}$ is an integral. It is in fact the linear combination of the integral $\mathcal{J}_1$ mentioned in equation 3 and of the time-dependent integral

$$\mathcal{T}_1' = t(P_1 + P_2 + P_3) - (p_1 + p_2 + p_3).$$

Being an integral and, at the same time, the infinitesimal generator of the group $\mathcal{L}$, the function $\mathcal{W}$ is the *momentum mapping* associated with the group $\mathcal{L}$. There follows that the group $\mathcal{L}$ may be used to reduce the three-particle Toda lattice to a Hamiltonian system with only two degrees of freedom.

FIRST REDUCTION

Classical mechanics handles the reduction of $\mathcal{H}$ commanded by the momentum mapping by means of a canonical change of variables that sets the integral $\mathcal{J}_1$ as a momentum whose conjugate coordinate is to be ignorable in $\mathcal{H}$. In the present case, one might further require that the reducing transformation be a representation of the group $\mathcal{S}_3$ of discrete symmetries; in this manner, the reduction will at the same time propagate the symmetries of $\mathcal{H}$ to the transformed Hamiltonian. The reducing transformation we propose is a canonical extension of the point transformation.

$$(b, r, \phi) \mapsto (p_1, p_2, p_3) : \mathbf{R}^2 \times \mathbf{S}^1 \mapsto \mathbf{R}^3$$

defined by the equations

$$p_1 = b + r \cos \phi,$$
$$p_2 = b + r \cos (\phi + \tfrac{2}{3}\pi),$$
$$p_3 = b + r \cos (\phi - \tfrac{2}{3}\pi), \tag{12}$$

To the elements in the group $\mathcal{S}_3$ correspond the geometric rotations and inversions mentioned below:

$$t_0: (b, r, \phi) \mapsto (b, r, \phi);$$
$$t_1: (b, r, \phi) \mapsto (b, r, \phi + \tfrac{2}{3}\pi);$$
$$t_2: (b, r, \phi) \mapsto (b, r, \phi - \tfrac{2}{3}\pi);$$
$$t_3: (r, b, \phi) \mapsto (-b, -r, -\phi - \tfrac{2}{3}\pi);$$
$$t_4: (b, r, \phi) \mapsto (-b, -r, -\phi);$$
$$t_5: (b, r, \phi) \mapsto (-b, -r, -\phi + \tfrac{2}{3}\pi).$$

The transformation is most easily inverted after one has observed that

$$p_2 - p_3 = -r\sqrt{3}\sin\phi,$$
$$p_3 - p_1 = -r\sqrt{3}\sin(\phi + \tfrac{2}{3}\pi),$$
$$p_1 - p_2 = -r\sqrt{3}\sin(\phi - \tfrac{2}{3}\pi). \tag{13}$$

It is then readily seen that

$$b = \tfrac{1}{3}(p_1 + p_2 + p_3),$$
$$r = \pm \tfrac{1}{3}\sqrt{2[(p_1 - p_2)^2 + (p_2 - p_3)^2 + (p_3 - p_1)^2]}. \tag{14}$$

On account of the symmetry with respect to the group $\mathcal{S}_3$, either sign will do for r. Having computed the determinant

$$\frac{\partial(p_1, p_2, p_3)}{\partial(b, r, \phi)} = -\frac{3}{2}r\sqrt{3},$$

one realizes that the transformation presents a polelike singularity at $r = 0$, which, according to equation 14, corresponds to the line $p_1 = p_2 = p_3$. The angle ϕ is not defined at that point; anywhere else it should be computed by using the formulas

$$\tfrac{27}{4}r^3 \cos 3\phi = (2p_1 - p_2 - p_3)(2p_2 - p_3 - p_1)(2p_3 - p_1 - p_2),$$
$$\tfrac{3}{4}\sqrt{3}\,r^3 \sin 3\phi = (p_1 - p_2)(p_2 - p_3)(p_3 - p_1).$$

The coordinate transformation is extended into a *gauge-free* canonical transformation

$$(b, r, \phi, B, R, \Phi) \mapsto (p_1, p_2, p_3, P_1, P_2, P_3)$$

in the usual way by finding the momenta B, R, and Φ to satisfy exactly the differential identity

$$B\,db + R\,dr + \Phi\,d\phi = P_1\,dp_1 + P_2\,dp_2 + P_3\,dp_3.$$

Hence the implicit linear relations

$$B = P_1 \frac{\partial p_1}{\partial b} + P_2 \frac{\partial p_2}{\partial b} + P_3 \frac{\partial p_3}{\partial b},$$

$$R = P_1 \frac{\partial p_1}{\partial r} + P_2 \frac{\partial p_2}{\partial r} + P_3 \frac{\partial p_3}{\partial r},$$

$$\Phi = P_1 \frac{\partial p_1}{\partial \phi} + P_2 \frac{\partial p_2}{\partial \phi} + P_3 \frac{\partial p_3}{\partial \phi}.$$

These equations are solved to produce the explicit equations for the momenta:

$$P_1 = \frac{1}{3}\left(B + 2R \cos \phi - 2\frac{\Phi}{r} \sin \phi\right),$$

$$P_2 = \frac{1}{3}\left[B + 2R \cos\left(\phi + \frac{2}{3}\pi\right) - 2\frac{\Phi}{r} \sin\left(\phi + \frac{2}{3}\pi\right)\right],$$

$$P_3 = \frac{1}{3}\left[B + 2R \cos\left(\phi - \frac{2}{3}\pi\right) - 2\frac{\Phi}{r} \sin\left(\phi - \frac{2}{3}\pi\right)\right]. \tag{15}$$

Like in the case of the coordinates, inversion of the transformation is more expeditiously handled after one has calculated the differences

$$P_2 - P_3 = -\frac{2}{3}\sqrt{3}\left[R \sin \phi + \frac{\Phi}{r} \cos \phi\right],$$

$$P_3 - P_1 = -\frac{2}{3}\sqrt{3}\left[R \sin\left(\phi + \frac{2}{3}\pi\right) + \frac{\Phi}{r} \cos\left(\phi + \frac{2}{3}\pi\right)\right],$$

$$P_1 - P_2 = -\frac{2}{3}\sqrt{3}\left[R \sin\left(\phi + \frac{2}{3}\pi\right) + \frac{\Phi}{r} \cos\left(\phi - \frac{2}{3}\pi\right)\right].$$

It is then relatively easy to express the new momenta explicitly in terms of the old coordinates and momenta:

$$B = P_1 + P_2 + P_3,$$

$$3rR = (p_1 - p_2)(P_1 - P_2) + (p_2 - p_3)(P_2 - P_3) + (p_3 - p_1)(P_3 - P_1),$$

$$9\Phi^2 = [(p_2 - p_3)(P_3 - P_1) - (p_3 - p_1)(P_2 - P_3)]^2$$
$$+ [(p_3 - p_1)(P_1 - P_2) - (p_1 - p_2)(P_3 - P_1)]^2$$
$$+ [(p_1 - p_2)(P_2 - P_3) - (p_2 - p_3)(P_1 - P_2)]^2.$$

As intended, the canonical extension performs the reduction based on the momentum mapping $\mathcal{W}$. Indeed,

(a) It expresses the potential energy

$$\mathcal{V} = e^{-r\sqrt{3} \sin \phi} + e^{-r\sqrt{3} \sin (\phi + 2\pi/3)} + e^{-r\sqrt{3} \sin (\phi - 2\pi/3)} \tag{16}$$

as a function of r and ϕ only.

(b) It splits Hamiltonian equation 2 into the sum

$$\mathcal{K} = \tfrac{1}{6}B^2 + \mathcal{H},$$

the second term in the sum being the function

$$\mathcal{H} = \frac{1}{3}\left(R^2 + \frac{\Phi^2}{r^2}\right) + \mathcal{V}.$$

The coordinate b and its momentum B are both ignorable in $\mathcal{H}$. Therefore the canonical equations separate into a four-dimensional system

$$\dot{r} = \frac{\partial \mathcal{H}}{\partial R}, \qquad \dot{\phi} = \frac{\partial \mathcal{H}}{\partial \Phi}, \qquad \dot{R} = -\frac{\partial \mathcal{H}}{\partial r}, \qquad \dot{\Phi} = -\frac{\partial \mathcal{H}}{\partial \phi}$$

and a quadrature

$$\dot{b} = \frac{\partial \mathcal{H}}{\partial B} = \frac{1}{3} B.$$

The reduction induced by the momentum mapping $(p, P, t) \mapsto \mathcal{W}$ does not alter the fact that the system is integrable. Therefore the reduced system admits a second integral. Clearly, it must be a function of $\mathcal{I}_1 = B$, $\mathcal{I}_2$, and $\mathcal{I}_3$. One would be well advised to choose as a representative of the second integral a function in which B and b are both ignorable. A candidate is

$$\mathcal{J} = (2P_1 - P_2 - P_3)(2P_2 - P_3 - P_1)(2P_3 - P_1 - P_2)$$
$$- 9[(2P_1 - P_2 - P_3)e^{p_2-p_3} + (2P_2 - P_3 - P_1)e^{p_3-p_1}$$
$$+ (2P_3 - P_1 - P_2)e^{p_1-p_2}]$$

In terms of Hénon's integrals,

$$\mathcal{J} = 2\mathcal{I}_1^3 + 27\mathcal{I}_3 - 9\mathcal{I}_1\mathcal{I}_2.$$

In the reducing coordinates just introduced,

$$\mathcal{J} = 2\left[R^3 \cos 3\phi - 3R^2\frac{\Phi}{r}\sin 3\phi - 3R\left(\frac{\Phi}{r}\right)^2 \cos 3\phi + \left(\frac{\Phi}{r}\right)^3 \sin 3\phi\right]$$
$$- 18\left(R\cos\phi - \frac{\Phi}{r}\sin\phi\right)e^{-r\sqrt{3}\sin\phi}$$
$$- 18\left[R\cos\left(\phi + \frac{2}{3}\pi\right) - \frac{\Phi}{r}\sin\left(\phi + \frac{2}{3}\pi\right)\right]e^{-r\sqrt{3}\sin[\phi+(2/3)\pi]}$$
$$- 18\left[R\cos\left(\phi - \frac{2}{3}\pi\right) - \frac{\Phi}{r}\sin\left(\phi - \frac{2}{3}\pi\right)\right]e^{-r\sqrt{3}\sin[\phi-(2/3)\pi]}.$$

EXPANSION AT THE EQUILIBRIUM

The reduction maps the class of solutions

$$p_1 = p_2 = p_3 = b, \qquad P_1 = P_2 = P_3 = \frac{1}{3}B$$

onto the origin in the reduced phase space; this explains why the origin turns out to be an equilibrium for the reduced Toda lattice. From here on, the analysis will be confined to a neighborhood of that equilibrium. Moreover, only nonnegative values will be retained for r; by virtue of this restriction, the symmetries t_3, t_4, and t_5 are forbidden, and the discrete group of symmetries $\mathcal{S}_6$ is pared down to the subgroup $\mathcal{D}_3$ of rotations by $2\pi/3$ about the origin.

The potential energy equation 16 is expanded at the origin as a series

$$\mathcal{V} = \sum_{n \geq 0} r^n \mathcal{V}_n(\phi)$$

in the powers of r. On account of the symmetry with respect to the group $\mathcal{D}_3$, for any $n \geq 0$, V_n is periodic in ϕ with period $2\pi/3$. A few elementary commands in MACSYMA produce the following terms in the expansion of $\mathcal{V}$:

$$\mathcal{V}_0 = 3, \qquad \mathcal{V}_6 = \frac{9}{2560}(10 - \cos 6\phi),$$

$$\mathcal{V}_1 = 0, \qquad \mathcal{V}_7 = \frac{27\sqrt{3}}{5120}\sin 3\phi,$$

$$\mathcal{V}_2 = \frac{9}{4}, \qquad \mathcal{V}_8 = \frac{27}{573440}(35 - 8\cos 6\phi),$$

$$\mathcal{V}_3 = \frac{3\sqrt{3}}{8}\sin 3\phi, \qquad \mathcal{V}_9 = \frac{3\sqrt{3}}{1146880}(84\sin 3\phi - \sin 9\phi),$$

$$\mathcal{V}_4 = \frac{27}{64}, \qquad \mathcal{V}_{10} = \frac{81}{22937600}(14 - 5\cos 6\phi),$$

$$\mathcal{V}_5 = \frac{9\sqrt{3}}{128}\sin 3\phi, \qquad \ldots$$

By dropping the term of degree zero from $\mathcal{H}$, we set the energy at the equilibrium as the zero point on the energy scale. As is expected, the term of first degree in r vanishes. Hence, in the neighborhood of the origin, the Hamiltonian $\mathcal{H}$ begins with the term

$$\mathcal{T}_0 = \frac{1}{3}\left(R^2 + \frac{\Phi^2}{r^2}\right) + r^2\mathcal{V}_2 = \frac{1}{3}\left(R^2 + \frac{\Phi^2}{r^2}\right) + \frac{9}{4}r^2.$$

As most authors have done, we rescale the radial distance r and its conjugate momentum R. While we are at it, we also reorient the coordinate plane so as to conform to the coordinates generally adopted in the most recent literature. All these modifications are implemented via the *gauge-free* canonical transformation

$$r = \sqrt{\frac{2}{3}}r', \qquad \phi = \phi' - \frac{\pi}{2}, \qquad R = \sqrt{\frac{3}{2}}R', \qquad \Phi = \Phi'.$$

After this transformation,

$$\mathcal{T}_0 = \frac{1}{2}\left(R'^2 + \frac{\Phi'^2}{r'^2}\right) + \frac{3}{2}r'^2.$$

It represents the Hamiltonian of an elliptic oscillator; that is to say, a pair of harmonic oscillators whose natural frequency is, for each of them, equal to $\sqrt{3}$.

The momentum Φ' is the angular momentum of the elliptic oscillator; since ϕ' is ignorable in $\mathcal{T}_0$, Φ' is an integral of that system. Starting with Φ' as the principal term at order zero, Whittaker's algorithm for constructing an adelphic integral, possibly with modifications of the type proposed by Contopoulos to circumvent the difficulties created by the $1-1$ resonance, should aim at generating a second integral for the Toda lattice in the neighborhood of the origin. However, without going through that procedure, we already know a second integral of the reduced Toda lattice beginning with Φ' at $r' = 0$, namely the function $\mathcal{J}$. Authors who choose to create second or third integrals by Whittaker's algorithm should tell us how their contrived integrals relate to $\mathcal{J}$.

Following in the footsteps of Contopoulos and Polymillis,[12] we propose to analyze the successive truncations

$$\mathcal{T}_n = \mathcal{T}_{n-1} + \epsilon^n r^{n+2} \mathcal{V}_{n+2} \qquad \text{for } n \geq 1.$$

It has long been noticed that the first-order truncation

$$\mathcal{T}_1 = \mathcal{T}_0 + \epsilon r^3 \mathcal{V}_3 = \mathcal{T}_0 + \tfrac{1}{4}\sqrt{2}\epsilon r^3 \cos 3\phi'$$

represents the Hénon-Heiles oscillator.

SECOND REDUCTION

Perturbed elliptic oscillators are most easily dealt with in the Lissajous map (l, g, L, G).

The Lissajous variables are closely related to the ellipses that are the trajectories in the dynamical system described by the Hamiltonian $\mathcal{T}_0$. Those ellipses have the origin as their center of symmetry. Let a and b denote, respectively, the semimajor and the semiminor axis. Let X be the position at a given instant of a particle moving on the circle of radius a centered at the origin; assume that X revolves at the constant angular velocity n, and call l the longitude of X reckoned from the semiminor axis. [N.B. In the problem at hand, $n = \sqrt{3}$.] Let Y stand for the affine projection of X onto the ellipse; if f is the longitude of Y also reckoned from the semiminor axis, then

$$r' \cos f = b \cos l \quad \text{and} \quad r' \sin f = a \sin l.$$

Let g be the longitude of the semiminor axis reckoned from the direction from which ϕ' is measured. Evidently,

$$\phi' = g + f;$$

then there follows that

$$r' \cos \phi' = r \cos f \cos g - r \sin f \sin g$$
$$= \frac{1}{2}(a+b)\cos(l+g) - \frac{1}{2}(a-b)\cos(l-g),$$
$$r' \sin \phi' = r \cos f \sin g + r \sin f \cos g$$
$$= \frac{1}{2}(a+b)\sin(l+g) + \frac{1}{2}(a-b)\sin(l-g).$$

With the angles l and g are associated the momenta

$$L = \tfrac{1}{2}n(a^2 + b^2) \quad \text{and} \quad G = nab.$$

The definition of the Lissajous transformation as a one-parameter canonical mapping

$$\mathcal{K}:(l, g, L, G; n) \mapsto (r', \phi', R', \Phi'),$$

is completed by postulating that

$$R' = n\frac{\partial r'}{\partial l} = \frac{1}{r'}(a^2 - b^2)\sin l \cos l \quad \text{and} \quad \Phi' = G.$$

Expressed in terms of the Lissajous variables, the Hamiltonian $\mathcal{T}_0$ of the elliptic oscillator reduces to the function

$$\mathcal{T}_0 = nL,$$

and the Lie derivative

$$\mathcal{L}_0 : f \mapsto \mathcal{L}_0(f) = (f; \mathcal{T}_0)$$

in the flow of the elliptic oscillator takes the form of a single partial derivative

$$\mathcal{L}_0 = n\frac{\partial}{\partial l}. \tag{17}$$

In that expression lies the main reason why perturbed elliptic oscillators are best handled in Lissajous variables. Normalization, we have explained earlier, consists in changing a perturbed Hamiltonian like $\mathcal{H}$ into a function $\mathcal{N}$ such that $\mathcal{L}_0(\mathcal{N}) = 0$, that is, such that $\mathcal{N} \in \ker(\mathcal{L}_0)$. But, according to equation 17, the kernel $\ker(\mathcal{L}_0)$ consists of those functions that do not depend on the *elliptic* anomaly l. Furthermore, because all the solutions of the elliptic oscillator are periodic, the space of perturbation functions may be decomposed as the direct sum of $\ker(\mathcal{L}_0)$ and of its image, $\text{im}(\mathcal{L}_0)$, which is the set of functions Φ for which there exists a function Ψ such that $\mathcal{L}_0(\Psi) = \Phi$. Again in view of equation 17, decomposing a perturbation function into its component in $\ker(\mathcal{L}_0)$ and its component in $\text{im}(\mathcal{L}_0)$ is very simple: if Φ is periodic in l, then its average $\langle \Phi \rangle_l$ over the elliptic anomaly belongs to $\ker(\mathcal{L}_0)$, and the difference $\Phi - \langle \Phi \rangle_l$ belongs to $\text{im}(\mathcal{L}_0)$. Moreover, any function of the form

$$\Psi = \int^l (\Phi - \langle \Phi \rangle_l)\, dl + \Omega \quad \text{with} \quad \Omega \in \ker(\mathcal{L}_0)$$

is such that

$$\mathcal{L}_0(\Psi) = \Phi - \langle \Phi \rangle_l.$$

Thus, by elementary manipulations of Fourier series in the elliptic anomaly rather than by intricate combinations of homogeneous polynomials based on the Theory of Invariants, can one achieve the normalization of a wide class of perturbed elliptic oscillators. The reduced three-particle Toda lattice belongs evidently to that class.

When expressed in terms of the Lissajous variables, the term $r^n \mathcal{V}_n$ appears as a homogeneous polynomial of degree n in the half-sum and half-difference

$$\sigma = \frac{1}{2}(a+b) = \sqrt{\frac{(L+G)}{2n}}, \quad \delta = \frac{1}{2}(a-b) = \sqrt{\frac{(L-G)}{2n}},$$

the polynomial coefficients being sums of cosines in integer combinations of the angles l and g. For the three-particle Toda lattice, for instance, upon setting that

$$\mathcal{H} = \sum_{n \geq 0} \frac{1}{n!} \epsilon^n \mathcal{H}_n,$$

one finds for the first few orders that

$$\mathcal{H}_0 = L\sqrt{3},$$

$$\mathcal{H}_1 = \frac{\sqrt{2}}{4} [3\sigma\delta^2 \cos(l - 3g) - 3\sigma^2\delta \cos(l + 3g)$$

$$- \delta^3 \cos 3(l - g) + \sigma^3 \cos 3(l + g)],$$

$$\mathcal{H}_2 = \frac{3}{8} [\sigma^4 + 4\sigma^2\delta^2 + \delta^4 - 4\sigma\delta(\sigma^2 + \delta^2)\cos 2l + 2\sigma^2\delta^2 \cos 4l],$$

$$\mathcal{H}_3 = \frac{3\sqrt{2}}{16} [\sigma\delta^2(6\sigma^2 + 4\delta^2)\cos(l - 3g) - \sigma^2\delta(4\sigma^2 + 6\delta^2)\cos(l + 3g)$$

$$- \delta^3(4\sigma^2 + \delta^2)\cos 3(l - g) + \sigma^3(\sigma^2 + 4\delta^2)\cos 3(l + g)$$

$$+ \sigma\delta^4 \cos(5l - 3g) - \sigma^4\delta \cos(5l + 3g)],$$

$$\mathcal{H}_4 = \frac{1}{40} [10(\sigma^6 + 9\sigma^4\delta^2 + 9\sigma^2\delta^4 + \delta^6) - 60\sigma\delta(\sigma^4 + 3\sigma^2\delta^2 + \delta^4)\cos 2l$$

$$+ 60\sigma^2\delta^2(\sigma^2 + \delta^2)\cos 4l - 20\sigma^3\delta^3 \cos 6l$$

$$+ 15\sigma^2\delta^4 \cos(2l - 6g) + 15\sigma^4\delta^2 \cos(2l + 6g)$$

$$- 6\sigma\delta^5 \cos(4l - 6g) - 6\sigma^5\delta \cos(4l + 6g)$$

$$+ \sigma^6 \cos 6(l + g) + \delta^6 \cos 6(l - g) - 20\sigma^3\delta^3 \cos 6g].$$

Symmetry relative to the group $\mathcal{D}_3$ is reflected in the fact that the argument of perigee enters $\mathcal{H}$ in multiples of $3g$ only.

It can be proved that the normalized Hamiltonian will be a series of the form

$$\mathcal{N} = \sum_{n\geq 0} \frac{1}{2n!} \epsilon^{2n} \mathcal{N}_{2n},$$

whose terms will be homogeneous polynomials of degree $2m$ in σ and δ, with sums of cosines in $6g$ as coefficients.

The calculation of the normalizing Lie transformation was performed by MAO, a processor in LISP for *M*echanized *A*lgebraic *O*perations.

The list below reproduces the terms in the normalizations $\mathcal{T}'_n$ of the successive truncations $\mathcal{T}_n$, as we have defined them in equation 8. Note that, for any n, $\mathcal{N}_{2n}$ is identical to $\mathcal{T}_{2n,2n}$.

Degree 2:

$$\mathcal{T}_0 = L\sqrt{3};$$

Degree 3:

$$\mathcal{T}_{1,2} = \frac{1}{24} L^2 \left(1 - \frac{7}{2} e^2\right);$$

$$\mathcal{T}_{1,4} = L^3 \sqrt{3} \left(-\frac{11}{432} + \frac{7}{1152} e^2 + \frac{7}{144} e^3 \cos 6g\right);$$

Degree 4:

$$\mathcal{T}_{2,2} = \frac{1}{6} L^2 \left(1 - \frac{1}{2} e^2\right);$$

$$\mathcal{T}_{2,4} = L^3 \sqrt{3} \left(-\frac{17}{108} + \frac{11}{36} e^2 - \frac{1}{24} e^3 \cos 6g\right);$$

Degree 5:

$$\mathcal{T}_{3,4} = L^3 \sqrt{3} \left(-\frac{11}{108} + \frac{1}{18} e^2 + \frac{1}{144} e^3 \cos 6g\right);$$

Degree 6:

$$\mathcal{T}_{4,4} = L^3 \sqrt{3} \left(-\frac{2}{27} + \frac{7}{72} e^2\right);$$

$$\mathcal{T}_{4,6} = L^4 \left(\frac{2845}{5184} - \frac{5365}{5184} e^2 + \frac{19135}{41472} e^4 - \frac{5}{24} e^3 \cos 6g\right);$$

Degree 7:

$$\mathcal{T}_{5,6} = L^4 \left(\frac{3385}{5184} - \frac{8065}{5184} e^2 + \frac{12115}{41472} e^4 + \frac{5}{192} e^3 \cos 6g\right);$$

Degree 8:

$$T_{6,6} = L^4 \left(\frac{55}{81} - \frac{1915}{1296} e^2 + \frac{1565}{5184} e^4 \right);$$

$$T_{6,8} = L^5 \sqrt{3} \left[-\frac{317947}{58320} + \frac{393491}{23328} e^2 - \frac{39851}{5832} e^4 + \left(\frac{5033}{1728} - \frac{77}{24} e^2 \right) e^3 \cos 6g \right];$$

Degree 9:

$$T_{7,8} = L^5 \sqrt{3} \left[-\frac{314167}{58320} + \frac{385931}{23328} e^2 - \frac{167909}{23328} e^4 + \left(\frac{5467}{1728} - \frac{21679}{6912} e^2 \right) \right.$$
$$\left. \cdot e^3 \cos 6g \right];$$

Degree 10:

$$T_{8,8} = L^5 \sqrt{3} \left[-\frac{3920}{729} + \frac{387065}{23328} e^2 - \frac{669935}{93312} e^4 + \frac{5425}{1728} (1 - e^2) e^3 \cos 6g \right];$$

$$T_{8,10} = L^6 \left[\frac{11372963}{46656} - \frac{30578975}{31104} e^2 + \frac{9276995}{13824} e^4 - \frac{48890905}{746496} e^6 \right.$$
$$\left. + \left(-\frac{24928855}{55296} + \frac{24862495}{55296} e^2 \right) e^3 \cos 6g + \frac{175}{18432} e^6 \cos 12g \right];$$

Degree 11:

$$T_{9,10} = L^6 \left[\frac{11378633}{46656} - \frac{30595985}{31104} e^2 + \frac{9256205}{13824} e^4 - \frac{48898635}{746496} e^6 \right.$$
$$\left. - \left(-\frac{24891895}{55296} + \frac{24896515}{55296} e^2 \right) e^3 \cos 6g - \frac{7}{18432} e^6 \cos 12g \right];$$

Degree 12:

$$T_{10,10} = L^6 \left[\frac{177800}{729} - \frac{15296575}{15552} e^2 + \frac{4628575}{6912} e^4 - \frac{6124475}{93312} e^6 - \frac{24895255}{55296} \right.$$
$$\left. \cdot (1 - e^2) e^3 \cos 6g \right];$$

$$T_{10,12} = L^7 \sqrt{3} \left[-\frac{2945998495}{489888} + \frac{8441432131}{279936} e^2 - \frac{31929190865}{1119744} e^4 \right.$$
$$+ \frac{14116306655}{2239488} e^6 + \left(\frac{21345584689}{995328} - \frac{25149662351}{995328} e^2 \right.$$
$$\left. \left. + \frac{1901248217}{497664} e^4 \right) e^3 \cos 6g + \frac{319}{9216} e^6 \cos 12g \right];$$

Degree 13:

$$\mathcal{T}_{11,12} = L^7 \sqrt{3} \left[-\frac{2945961073}{489888} + \frac{8441357287}{279936} e^2 - \frac{31931061965}{1119744} e^4 \right.$$
$$+ \frac{14115183995}{2239488} e^6 + \left(\frac{21346107409}{995328} - \frac{25148605031}{995328} e^2 \right.$$
$$\left. \left. + \frac{950645641}{248832} e^4 \right) e^3 \cos 6g - \frac{11}{9216} e^6 \cos 12g \right];$$

Degree 14:

$$\mathcal{T}_{12,12} = L^7 \sqrt{3} \left[-\frac{13151600}{2187} + \frac{8441373325}{279936} e^2 - \frac{31930981775}{1119744} e^4 \right.$$
$$+ \frac{14115210725}{2239488} e^6 + \left(\frac{21346059889}{995328} - \frac{25148640671}{995328} e^2 \right.$$
$$\left. \left. + \frac{1901290391}{497664} e^4 \right) e^3 \cos 6g \right];$$

$$\mathcal{T}_{12,14} = L^8 \left[\frac{26565335852627}{40310784} - \frac{80080795649855}{20155392} e^2 \right.$$
$$+ \frac{1560555744452705}{322486272} e^4 - \left(\frac{1077583815743635}{644972544} e^6 \right.$$
$$+ \frac{292630636162865}{5159780352} e^8 + \left(-\frac{1615518571385719}{429981696} \right.$$
$$+ \frac{1050288049007197}{191102976} e^2 - \frac{2990528881565225}{1719926784} e^4 \right) e^3 \cos 6g$$
$$\left. + \left(-\frac{5442236342543}{31850496} + \frac{5442236342543}{31850496} e^2 \right) e^6 \cos 12g \right];$$

Degree 15:

$$\mathcal{T}_{13,14} = L^8 \left[\frac{26565341690459}{40310784} - \frac{80080801487687}{20155392} e^2 + \frac{1560554168238065}{322486272} e^4 \right.$$
$$- \frac{1077585600606515}{644972544} e^6 + \frac{292629906422865}{5159780352} e^8$$
$$+ \left(-\frac{1615518088791607}{429981696} + \frac{1050288870627997}{191102976} e^2 \right.$$
$$\left. - \frac{29990527025134649}{1719926784} e^4 \right) e^3 \cos 6g - \frac{5442236342543}{31850496}$$
$$\left. \cdot (1 - e^2) e^6 \cos 12g \right];$$

Degree 16:

$$T_{14,14} = L^8 \left[\frac{12971358400}{19683} - \frac{20020199824625}{5038848} e^2 + \frac{1560554233913675}{322486272} e^4 \right.$$

$$- \frac{1077585556822775}{644972544} e^6 + \frac{3657873972475}{644972544} e^8$$

$$+ \left(\frac{1615518135494263}{429981696} + \frac{105028844682077}{191102976} e^2 \right.$$

$$\left. - \frac{2990527060161641}{1719926784} e^4 \right) e^3 \cos 6g - \frac{5452243333527}{31850496}$$

$$\left. \cdot (1 - e^2) e^6 \cos 12g \right];$$

$$T_{14,16} = L^9 \sqrt{3} \left[- \frac{4549314831078875}{136048896} + \frac{7152213180654685}{30233088} e^2 \right.$$

$$- \frac{2273052824181189905}{644972544} e^4 + \frac{19121791731291499225}{11609505792} e^6$$

$$- \frac{56520991589487425}{3869835264} e^8 + \left(\frac{352333651535622185}{1289945088} \right.$$

$$\left. - \frac{2455604246371046285}{5159780352} e^2 + \frac{4335101517913888865}{5159780352} e^4 \right.$$

$$\left. - \frac{150023524298950445}{20639121408} e^6 \right) e^3 \cos 6g + \frac{3835506038493840405}{10319450704}$$

$$\left. \cdot (1 - e^2) e^6 \cos 12g - \frac{5005}{5971968} e^9 \cos 18g \right];$$

Degree 17:

$$T_{15,16} = L^9 \sqrt{3} \left[- \frac{4549314848569235}{136048896} + \frac{7152213180654685}{30233088} e^2 \right.$$

$$- \frac{2273052852202782650}{644972544} e^4 + \frac{19121791170859627250}{11609505792} e^6$$

$$- \frac{56520994216511825}{3889835264} e^8 + \left(\frac{352333652469675305}{1289945088} \right.$$

$$\left. - \frac{2455604120684585585}{5159780352} e^2 + \frac{4335101571154916705}{20639121408} e^4 \right.$$

$$\left. - \frac{150023522197330925}{20639121408} e^6 \right) e^3 \cos 6g + \left(\frac{383506034757627925}{10309450704} \right.$$

$$\left. - \frac{383506034883650965}{10319560704} e^2 \right) e^6 \cos 12g + \frac{715}{47775744} e^9 \cos 18g \right];$$

Degree 18:

$$\mathcal{T}_{16,16} = L^9\sqrt{3}\left[-\frac{17770761008000}{531441} + \frac{7152213182739625}{30230088}e^2\right.$$
$$-\frac{227305285103521625}{644972544}e^4 + \frac{1912179118253529125}{11609505792}e^6$$
$$-\frac{56520994180025375}{3869835264}e^8 + \left(\frac{352333652365891625}{1289945088}\right.$$
$$-\frac{2455604121462963185}{5159780352}e^2 + \frac{4335101570220863585}{206391121408}e^4$$
$$\left.-\frac{15002352223276845}{20639121408}e^6\right)e^3\cos 6g$$
$$\left.+\frac{383506034861411605}{10319560704}(1-e^2)e^6\cos 12g\right];$$

$$\mathcal{T}_{16,18} = L^{10}\left[\frac{55104745499762109437}{8162933760} - \frac{1796417306275776228297}{3265173504}e^2\right.$$
$$+\frac{323510426384894081099767}{3343537668096}e^4 - \frac{3805043960004706092255}{6687085336192}e^6$$
$$+\frac{56343585977655257036176 5}{6687075336192}e^8 + \frac{321682409595305239481}{2089711042560}e^{10}$$
$$+\left(-\frac{1889425792926369704048 9}{247669456896} + \frac{3877289851901623514056 7}{247669456896}e^2\right.$$
$$\left.-\frac{18587224837202460761120 7}{1981355655168}e^4 + \frac{2684312360162471374235 9}{1981355655168}e^6\right)e^3\cos 6g$$
$$+\left(-\frac{11844577056148085306057}{660451885056} + \frac{28199505777264065348819}{1320903770112}e^2\right.$$
$$\left.\left.-\frac{4510351661639903500993}{1320903770112}e^4\right)e^6\cos 12g - \frac{75361}{5308416}e^9\cos 18g\right];$$

Degree 19:

$$\mathcal{T}_{17,18} = L^{10}\left[\frac{55104745501197590627}{8162933760} - \frac{1796417306261407471 07}{3265173504}e^2\right.$$
$$+\frac{323510426348880729005047}{3343537668096}e^4 - \frac{3805043961290897238463 55}{6687075338192}e^6$$
$$\left.+\frac{56343859406745749073 65}{6687075336192}e^8 + \frac{321682409283805821251}{2089711042560}e^{10}\right.$$

$$+ \left(-\frac{18894257928958822102121}{247669456896} + \frac{3877289852286528127463}{247669456896} e^2 \right.$$

$$\left. -\frac{18587224835443342668903}{1981355655168} e^4 + \frac{26843123604940228697111}{1981355655168} e^6 \right) e^3 \cos 6g$$

$$+ \left(-\frac{11844577057113522610889}{660451885056} + \frac{281995057758558334906}{1320903770112} e^2 \right.$$

$$\left. -\frac{4510351664702058106432 1}{1320903770112} e^4 \right) e^6 \cos 12g + \frac{2431}{106116832} e^9 \cos 18g \Bigg];$$

Degree 20:

$$\mathcal{T}_{18,18} = L^{10} \Bigg[\frac{10762645605712000}{1594323} - \frac{89820865312855051375}{1632596752} e^2$$

$$+ \frac{32351042635042415838053 5}{3343537668096} e^4 - \frac{3805049600047060922235 5}{6687075336192} e^8$$

$$+ \frac{5634358977652570361765}{6687075336192} e^8 + \frac{321682409595305239481}{2089711042560} e^{10}$$

$$+ \left(-\frac{18894257929263697040489}{247669456896} + \frac{38772898519016235140567}{247669456896} e^2 \right.$$

$$\left. -\frac{18587224838202460861120 7}{1981355655168} e^4 + \frac{26843123604902119329815}{1981355655168} e^6 \right) e^3 \cos 6g$$

$$+ \left(-\frac{11845770570881163660 25}{660451885058} + \frac{28199505775877610271955}{1320903770112} e^2 \right.$$

$$\left. -\frac{4510351661701377539905}{1320903770112} e^4 \right) e^6 \cos 12g \Bigg].$$

GLOBAL BEHAVIOR

Once the elliptic anomaly has been eliminated, the three-particle Toda lattice is reduced to only one degree of freedom. In the traditional language of Hamiltonian mechanics, the equations of motion are reduced to the system

$$\dot{g} = \frac{\partial \mathcal{T}'_n}{\partial G}, \qquad \dot{G} = -\frac{\partial \mathcal{T}'_n}{\partial g}$$

whose integration is to be followed by the quadrature

$$\dot{l} = \frac{\partial \mathcal{T}'_n}{\partial L}.$$

This analytical reduction has a geometric interpretation. We look at the reduced phase space as composed of layers, each one made of a manifold $L =$ constant. On each layer, starting at a given point Q of coordinates (l, g, L, G), we follow the curve made of the points obtained from Q by changing the coordinates of P into the variables $(l + \epsilon, g, L, G)$, where l ranges over the interval $0 \leq l < 2\pi$. Such a curve belongs to the manifold of P; it is made of an ellipse centered at the origin in the plane of positions, and of an ellipse, also centered at the origin, in the plane of velocities. This object is then treated as a point in a new phase space. Mathematicians prove that, for each value of the integral L, the collection of these objects makes a symplectic manifold—the reduced phase space; we label it $\mathcal{P}(L)$.

However abstract it may be, the concept of a reduced phase space leads to a very intuitive geometric picture. Each object in $\mathcal{P}(L)$ is unambiguously characterized by the functions

$$\xi_1 = \tfrac{1}{2}Le \cos 2g \qquad \xi_2 = \tfrac{1}{2}Le \sin 2g \qquad \xi_3 = \tfrac{1}{2}G.$$

They serve as coordinates on $\mathcal{P}(L)$. Since

$$\xi_1^2 + \xi_2^2 + \xi_3^2 = \tfrac{1}{4}L^2,$$

the reduced phase space $\mathcal{P}(L)$ appears as a sphere. The point $(\xi_1 = \xi_2 = 0, \xi_3 = \tfrac{1}{2}L)$, which we call the *north pole*, represents the circle of radius $a = (L/n)^{1/2}$ traveled in the direct sense since $G = L > 0$. Likewise, the point diametrically opposite, at $\xi_1 = \xi_2 = 0$, $\xi_3 = -\tfrac{1}{2}L$, is what we call the *south pole*, and it represents the same circular orbit as the north pole, but traveled in the retrograde sense. Points on the equator $\xi_3 = 0$ represent rectilinear orbits $(G = 0)$. Note that the variables (g, G) constitute only a local map on $\mathcal{P}(L)$: at the poles, the angle g is not determined.

The equations of the reduced system are easy to write in the spherical coordinates after one has calculated the Poisson brackets

$$(\xi_1, \xi_2) = \xi_3, \qquad (\xi_2, \xi_3) = \xi_1, \qquad (\xi_3, \xi_1) = \xi_2.$$

As a consequence, by virtue of Liouville's theorem, one finds, for instance, that

$$\dot{\xi}_1 = (\xi_1; T'_n) = (\xi_1, \xi_2)\frac{\partial T'_n}{\partial \xi_2} + (\xi_1, \xi_3)\frac{\partial T'_n}{\partial \xi_3} = \xi_3 \frac{\partial T'_n}{\partial \xi_2} - \xi_2 \frac{\partial T'_n}{\partial \xi_3}.$$

The global evolution of the reduced system will therefore be studied in the differential system

$$\dot{\xi}_1 = \xi_3 \frac{\partial T'_n}{\partial \xi_2} - \xi_2 \frac{\partial T'_n}{\partial \xi_3},$$

$$\dot{\xi}_2 = \xi_1 \frac{\partial T'_n}{\partial \xi_3} - \xi_3 \frac{\partial T'_n}{\partial \xi_1},$$

$$\dot{\xi}_3 = \xi_2 \frac{\partial T'_n}{\partial \xi_1} - \xi_1 \frac{\partial T'_n}{\partial \xi_2}.$$

For $1 \leq n \leq 18$, each normalized truncated Hamiltonian T'_n turns out to be a function of the quantities

$$A = L^2 e^2 = 4(\xi_1^2 + \xi_2^2), \qquad B = L^3 e^3 \cos 6g = 8\xi_1(\xi_1^2 - 3\xi_2^2).$$

Therefore, in the Toda system,

$$\frac{\partial T'_n}{\partial \xi_i} = \frac{\partial T'_n}{\partial A}\frac{\partial A}{\partial \xi_i} + \frac{\partial T'_n}{\partial B}\frac{\partial B}{\partial \xi_i};$$

in consequence, the reduced equations take the form

$$\dot{\xi}_1 = 8\xi_2\xi_3\left(\frac{\partial T'_n}{\partial A} - 6\xi_1\frac{\partial T'_n}{\partial B}\right),$$

$$\dot{\xi}_2 = -8\xi_3\left[\xi_1\frac{\partial T'_n}{\partial A} + 3(\xi_1^2 - \xi_2^2)\frac{\partial T'_n}{\partial B}\right],$$

$$\dot{\xi}_3 = 24\xi_2(3\xi_1^2 - \xi_2^2)\frac{\partial T'_n}{\partial B}.$$

It is now quite clear that the north and south poles are equilibria. Straightforward derivations by machine shows that these equilibria are stable for each normalized truncation ($1 \leq n \leq 18$).

Looking for equilibria on the orbital sphere elsewhere than at the poles, it is legitimate to look in the cylindrical variables (g, G). The right-hand members of the reduced differential equations ($1 \leq n \leq 18$)

$$\dot{g} = \frac{\partial T'_n}{\partial G}, \qquad \dot{G} = -\frac{\partial T'_n}{\partial g}$$

have been derived automatically by machine. In each case, it was found that, at $G = 0$, that is, in the equator on the orbital sphere $\mathcal{P}(L)$, $\dot{g} = 0$. Moreover, in the equator, $\dot{G} = -\partial \mathcal{N}_n/\partial g$ vanishes identically, whereas $\dot{G} = -\partial T'_{n,m}/\partial g$ for $m = 2(\lfloor n/2 \rfloor + 1)$ vanishes if and only if $g \equiv 0 \pmod{\pi/6}$. In other words, the whole equator is a locus of parabolic equilibria for the truncated Hamiltonian T_{2n} or T_{2n+1} until the normalization introduces the term T_{2n+2}, at which moment all points in the equator cease to be equilibria except those at the vertices of the hexagon such that $g \equiv 0 \pmod{\pi/6}$. In the words of Carles Simó, a truncated Toda system shakes off degeneracy in the equator *at the first opportunity,* more precisely, as soon as the normalization handles terms of a degree higher than that of the truncation.

The behavior of a normalized truncated Toda system is even more intriguing when one looks at the effect of the perturbation on the stability of the surviving equilibria. For each $1 \leq n \leq 17$, a computer procedure sets up the variational equations in literal form at the six isolated equilibria in the equator, and solves them to obtain the characteristic exponents. The systems being Hamiltonian with only one degree of freedom, the characteristic equations are all of the form $\lambda^2 = \alpha\epsilon^{2\lfloor n/2 \rfloor + 2} + \cdots$, which says that the stability of the equilibrium is decided by the sign of α. The equilibrium is stable when α is negative; unstable otherwise. The list of these characteristic equations may be long, but it deserves to be reproduced *in extenso* because it shows a very intriguing pattern.

Order 2 (degree 4)

at $2g \equiv 0 \pmod{2\pi/3}$: $\lambda^2 = -\dfrac{15}{4}\sqrt{3}\,\dfrac{\epsilon^6}{6!}L^3 + \mathcal{O}(\epsilon^7)$;

at $2g \equiv \pi \pmod{2\pi/3}$: $\lambda^2 = \dfrac{15}{4}\sqrt{3}\,\dfrac{\epsilon^6}{6!}L^3 + \mathcal{O}(\epsilon^7)$.

Order 3 (degree 5)

at $2g \equiv 0 \pmod{2\pi/3}$: $\lambda^2 = \dfrac{5}{8}\sqrt{3}\,\dfrac{\epsilon^6}{6!}L^3 + \mathcal{O}(\epsilon^7)$;

at $2g \equiv \pi \pmod{2\pi/3}$: $\lambda^2 = -\dfrac{5}{8}\sqrt{3}\,\dfrac{\epsilon^6}{6!}L^3 + \mathcal{O}(\epsilon^7)$.

Order 4 (degree 6)

at $2g \equiv 0 \pmod{2\pi/3}$: $\lambda^2 = -35\,\dfrac{\epsilon^8}{8!}L^4 + \mathcal{O}(\epsilon^9)$;

at $2g \equiv \pi \pmod{2\pi/3}$: $\lambda^2 = 35\,\dfrac{\epsilon^8}{8!}L^4 + \mathcal{O}(\epsilon^9)$.

Order 5 (degree 7)

at $2g \equiv 0 \pmod{2\pi/3}$: $\lambda^2 = \dfrac{35}{8}\,\dfrac{\epsilon^8}{8!}L^4 + \mathcal{O}(\epsilon^9)$;

at $2g \equiv \pi \pmod{2\pi/3}$: $\lambda^2 = -\dfrac{35}{8}\,\dfrac{\epsilon^8}{8!}L^4 + \mathcal{O}(\epsilon^9)$.

Order 6 (degree 8)

at $2g \equiv 0 \pmod{2\pi/3}$: $\lambda^2 = -\dfrac{2555}{32}\sqrt{3}\,\dfrac{\epsilon^{10}}{10!}L^5 + \mathcal{O}(\epsilon^{11})$;

at $2g \equiv \pi \pmod{2\pi/3}$: $\lambda^2 = \dfrac{2555}{32}\sqrt{3}\,\dfrac{\epsilon^{10}}{10!}L^5 + \mathcal{O}(\epsilon^{11})$.

Order 7 (degree 9)

at $2g \equiv 0 \pmod{2\pi/3}$: $\lambda^2 = \dfrac{945}{128}\sqrt{3}\,\dfrac{\epsilon^{10}}{10!}L^5 + \mathcal{O}(\epsilon^{11})$;

at $2g \equiv \pi \pmod{2\pi/3}$: $\lambda^2 = -\dfrac{945}{128}\sqrt{3}\,\dfrac{\epsilon^{10}}{10!}L^5 + \mathcal{O}(\epsilon^{11})$.

Order 8 (degree) 10

at $2g \equiv 0 \pmod{2\pi/3}$: $\lambda^2 = -\dfrac{58905}{128}\,\dfrac{\epsilon^{12}}{12!}L^6 + \mathcal{O}(\epsilon^{13})$;

at $2g \equiv \pi \pmod{2\pi/3}$: $\lambda^2 = \dfrac{62755}{128} \dfrac{\epsilon^{12}}{12!} L^6 + \mathcal{O}(\epsilon^{13})$.

Order 9 (degree 11)

at $2g \equiv 0 \pmod{2\pi/3}$: $\lambda^2 = \dfrac{2079}{64} \dfrac{\epsilon^{12}}{12!} L^6 + \mathcal{O}(\epsilon^{13})$;

at $2g \equiv \pi \pmod{2\pi/3}$: $\lambda^2 = -\dfrac{539}{16} \dfrac{\epsilon^{12}}{12!} L^6 + \mathcal{O}(\epsilon^{13})$.

Order 10 (degree 12)

at $2g \equiv 0 \pmod{2\pi/3}$: $\lambda^2 = -\dfrac{405405}{512} \sqrt{3} \dfrac{\epsilon^{14}}{14!} L^7 + \mathcal{O}(\epsilon^{15})$;

at $2g \equiv \pi \pmod{2\pi/3}$: $\lambda^2 = \dfrac{1448447}{1536} \sqrt{3} \dfrac{\epsilon^{14}}{14!} L^7 + \mathcal{O}(\epsilon^{15})$.

Order 11 (degree 13)

at $2g \equiv 0 \pmod{2\pi/3}$: $\lambda^2 = \dfrac{45045}{1024} \sqrt{3} \dfrac{\epsilon^{14}}{14!} L^7 + \mathcal{O}(\epsilon^{15})$;

at $2g \equiv \pi \pmod{2\pi/3}$: $\lambda^2 = -\dfrac{151151}{3072} \sqrt{3} \dfrac{\epsilon^{14}}{14!} L^7 + \mathcal{O}(\epsilon^{15})$.

Order 12 (degree 14)

at $2g \equiv 0 \pmod{2\pi/3}$: $\lambda^2 = -\dfrac{1936935}{512} \dfrac{\epsilon^{16}}{16!} L^8 + \mathcal{O}(\epsilon^{17})$;

at $2g \equiv \pi \pmod{2\pi/3}$: $\lambda^2 = \dfrac{7982975}{1536} \dfrac{\epsilon^{16}}{16!} L^8 + \mathcal{O}(\epsilon^{17})$.

Order 13 (degree 15)

at $2g \equiv 0 \pmod{2\pi/3}$: $\lambda^2 = \dfrac{173745}{1024} \dfrac{\epsilon^{16}}{16!} L^8 + \mathcal{O}(\epsilon^{17})$;

at $2g \equiv \pi \pmod{2\pi/3}$: $\lambda^2 = -\dfrac{216645}{1024} \dfrac{\epsilon^{16}}{16!} L^8 + \mathcal{O}(\epsilon^{17})$.

Order 14 (degree 16)

at $2g \equiv 0 \pmod{2\pi/3}$: $\lambda^2 = -\dfrac{23301135}{4096} \sqrt{3} \dfrac{\epsilon^{18}}{18!} L^9 + \mathcal{O}(\epsilon^{19})$;

at $2g \equiv \pi \pmod{2\pi/3}$: $\lambda^2 = \dfrac{339428375}{36884} \sqrt{3} \dfrac{\epsilon^{18}}{18!} L^9 + \mathcal{O}(\epsilon^{19})$.

Order 15 (degree 17)

at $2g \equiv 0 \pmod{2\pi/3}$: $\lambda^2 = \dfrac{6891885}{32768} \sqrt{3} \dfrac{\epsilon^{18}}{18!} L^9 + \mathcal{O}(\epsilon^{19})$;

at $2g \equiv \pi \pmod{2\pi/3}$: $\lambda^2 = -\dfrac{88476245}{294912} \sqrt{3} \dfrac{\epsilon^{18}}{18!} L^9 + \mathcal{O}(\epsilon^{19})$.

Order 16 (degree 18)

at $2g \equiv 0 \pmod{2\pi/3}$: $\lambda^2 = -\dfrac{804381435}{32768} \dfrac{\epsilon^{20}}{20!} L^{10} + \mathcal{O}(\epsilon^{21})$;

at $2g \equiv \pi \pmod{2\pi/3}$: $\lambda^2 = \dfrac{14015821105}{294912} \dfrac{\epsilon^{20}}{20!} L^{10} + \mathcal{O}(\epsilon^{21})$.

Order 17 (degree 19)

at $2g \equiv 0 \pmod{2\pi/3}$: $\lambda^2 = \dfrac{6235515}{8192} \dfrac{\epsilon^{20}}{20!} L^{10} + \mathcal{O}(\epsilon^{21})$;

at $2g \equiv \pi \pmod{2\pi/3}$: $\lambda^2 = -\dfrac{186834505}{147456} \dfrac{\epsilon^{20}}{20!} L^{10} + \mathcal{O}(\epsilon^{21})$.

For the normalized truncated Hamiltonians T'_{2n}, the vertices of the hexagon at $2g \equiv 0 \pmod{2\pi/3}$ are unstable, while those at $2g \equiv \pi \pmod{2\pi/3}$ are stable. Passing to the next degree of truncation makes the stability flip to instability, and conversely; for normalized truncated Hamiltonians T'_{2n+1}, the vertices of the hexagon at $2g \equiv \pi \pmod{2\pi/3}$ are unstable, while those at $2g \equiv 0 \pmod{2\pi/3}$ are stable.

CONCLUSIONS

The main liability of the approach developed in this paper is its limitation to manifolds of low energy in the neighborhood of the equilibrium at the origin after the first reduction.

Still the central point behind this effort is borne out, that any investigation of eventual stochasticity in the truncations of a three-particle Toda lattice must begin by considering normalization of low energies.

The great benefit of this research lies in the unveiling of the evanescent degeneracy along the manifold of orbits that, on the average, are rectilinear. As an application and example of the usefulness of this approach, the breakup of the degeneracy has been analyzed, with stability index calculated to a high order.

The phase flow of any truncation around the equilibrium at the origin is determined by the underlying orbits of the normalized system that are in turn governed by the *residual* beyond the degree at which truncation has been operated.

Considerable insight has been gained without integrating the equations of motion. Recent numerical explorations, admittedly at levels of energy beyond the reach of a classical normalization, resulted in startling conjectures about the integrability of some classes of truncations for the three-particle Toda lattice. A striking outcome of the local analysis undertaken here is the revelation of a mechanism by which, as the degree of the truncation tends to infinity, a family of nonintegrable truncations tends to adopt a behavior typical of an integrable system.

REFERENCES

1. TODA, M. 1967. Vibrations of a chain with nonlinear interaction. J. Phys. Soc. Jap. **22:** 431–436.
2. TODA, M. 1975. Studies of a non-linear lattice. Phys. Rep. **18:**1–124.
3. HENON, M. 1974. Integrals of the Toda lattice. Phys. Rev. B **9:**1921–1923.
4. FLASCHKA, H. 1974. The Toda lattice. II. Existence of integrals. Phys. Rev. B **9:**1924–1925.
5. LUNSFORD, G. & J. FORD. 1972. On the stability of periodic orbits for nonlinear oscillator systems in regions exhibiting stochastic behavior. J. Math. Phys. **13:**700–705.
6. YOSHIDA, H. 1987. Non-integrability of the fourth order truncated Toda Hamiltonian. Private communication.
7. YOSHIDA, H. 1987. Non-integrability of the truncated Toda lattice Hamiltonian at any orders. Private communication.
8. DEPRIT, A. 1969. Canonical transformation depending on a small parameter. Celest. Mech. **1:**12–30.
9. SIMÓ, C. 1988. Homoclinic and heteroclinic phenomena in some Hamiltonian systems. Presented at the Conf. on Hamlitonian Systems. The University of Colorado, Boulder, Colo., Aug. 1987. (The Proceedings will appear in Contemporary Mathematics, a collection edited by the American Mathematical Society.)
10. BRAUN, M. 1973. On the applicability of the third integral of motion. J. Differential Equations **13:**300–318.
11. CHURCHILL, R. C., M KUMMER & D. L. ROD. 1983. On averaging, reduction, and symmetry in Hamiltonian systems. J. Differential Equations **49:**359–414.
12. CONTOPOULOS, G. & C. POLYMILLIS. 1987. Approximations of the 3-particle Toda lattice. Physica **24D:**328–342.

Simplifications toward Integrability of Perturbed Keplerian Systems

SEBASTIAN FERRER[a] AND CAROL A. WILLIAMS[b,c]

[a]*Departmento de Física Teórica*
Universidad de Zaragoza
50009 Zaragoza, Spain

[b]*Department of Mathematics*
University of South Florida
Tampa, Florida 33620

INTRODUCTION

Few perturbed Keplerian systems are integrable. The underlying structure that precludes integrability is generally not revealed in an obvious way. Questions of integrability versus nonintegrability of a Hamiltonian system may be more easily answered in some variables than in others, since the underlying mathematical structure will manifest itself more clearly in some coordinate systems as opposed to others. Thus, in Hamiltonian dynamics, one looks to canonical transformations to shed light on the question of integrability. One of the more famous classes of canonical transformations, known as normalizations, actually reduces the number of degrees of freedom of a system while at the same time imparting to the transformed Hamiltonian $\mathcal{H}'$, a symmetry of the unperturbed system. If the variable eliminated is, say x, then $(X'; \mathcal{H}') = 0$. The transformed system possesses the integral X' and the symmetry that it generates. From this ideal situation, one should be able to repeat the process with the next degree of freedom and the next. For instance, when Brouwer succeeded by the method of Poincaré–von Zeipel with a Delaunay normalization of the artificial satellite theory, he continued with the elimination of the argument of perigee, while noting that this second normalization was not valid near the critical inclination. As an illustration of our opening remarks, note that the nonintegrability of the satellite theory at critical inclination is not obvious when the system is expressed in rectangular, polar–nodal, or even Delaunay variables. Yet after the Delaunay normalization, it is quite clear.

So successful has the Delaunay normalization been in perturbed Keplerian Hamiltonian systems that authors often claim its success as a demonstration of the system's integrability. Since it removes one degree of freedom from the system, its success is tantamount to having solved a conservative, two-degree-of-freedom system if the normalized Hamiltonian depends only on the momenta. Even if the dependence on the argument of the pericenter remains, its complexities are much more easily revealed in a normalized system where the trappings of the "short period" perturbations have been removed. With more degrees of freedom, the normalized Hamiltonian is more

---

[c]To whom correspondence should be addressed.

complex and it is difficult to generalize about what are the expected advantages of a normalization. But it is to be expected that such a technique would be of some value in reducing the complexity of the original problem.

When the Delaunay normalization does not work, as in planetary theory where small divisors preclude its success even at first order in, for example, the Neptune–Pluto problem, its failure is often taken as a sign of nonintegrability. Or, arguing, perhaps correctly, that this failure may be merely an artifact of the method of solution, astronomers seek new methods. Nevertheless, in the zonal satellite problem, the success of the Delaunay normalization was not questioned until Kozai had difficulty calculating second-order perturbations. Later, Coffey and Deprit[1,2] who sought to go to third and fourth order by applying a Delaunay normalization with a Lie transformation, were forced to stop their pursuits because of limitations of available computing power. Obviously to them, the normalization must either be considered hopeless or it would have to be simplified in some way if it were to lead to a solution of the satellite problem. The second alternative could be followed because Deprit[3] had found just such a simplification. The difficulties in these cases come not from questions regarding the mathematical convergence of the algorithms for normalization. Celestial Mechanics does not expect convergence; it has a history of successful applications of asymptotic expansions. But it does need to carry normalizations beyond first order if some reasonable degree of precision is to be reached. If the algorithm or the mathematical function to which it is applied is so cumbersome as to render low orders intractable, then indeed this normalization is "hopeless."

The quadratic Zeeman problem speaks particularly well for the need for high orders. The algorithm for its normalization is not particularly complex when a special set of variables, called the Lissajous variables,[4] is used. Nevertheless, the capacity of computers to generate the literal developments is exceeded after twelfth order. Yet Quantum Mechanics would like Celestial Mechanics to carry this normalization beyond order twelve. If this is to be done, the Zeeman problem will have to be further simplified with a method similar to the one used in the theory of artificial satellites.

And so we come to the conclusion that while normalization has solved many problems and has exposed the major mathematical difficulties in others, it is not *always* the best *first* approach to a problem. In this paper we address the more difficult problems of Celestial Mechanics that can still be treated as perturbed Keplerian systems. These include the artificial satellite theory and the lunar theory. We address the question of whether or not there may be transformations different from normalizations that might reveal the underlying structure of the perturbation. It is hoped in this way to see more clearly the questions as well as some of the answers still before us.

THE LUNAR THEORY

At present there is a need for an accurate lunar theory to provide an additional method for the determination of time. What is sought is a worldwide synchronization of clocks to within at least $\pm 10^{-9}$ sec. For some scientific efforts a determination up to three orders of magnitude better would be welcome. (For example, oceanographic surveys attempt to time the return of laser pulses off of artificial satellites with a precision of $\pm 10^{-11}$ sec.) This kind of global precision is achieved by the radio astronomers who, with very long baseline interferometery (VLBI) networks, observe quasars from several positions on the surface of the earth simultaneously. However, a backup method is desirable and may become essential if the VLBI networks are temporarily silenced. And so we follow history and look to the moon to provide us with a master clock. Using the cubic laser reflectors on the surface of the moon, placed there by astronauts, we could produce Uniform Time with a precision of $\pm 10^{-11}$ sec. if we had a theory giving the location of the center of mass of the moon together with the orientation of its axis of figure (the librations) and the selenographic coordinates of the reflectors all with a precision of, say, ± 0.3 cm.

Several lunar theories are currently in use around the world. For example, the Nautical Almanac Office of the U.S. Naval Observatory uses numerical integrations for the lunar ephemeris. The Bureau des Longitudes is currently taking the lunar ephemeris from ELP 2000.[5] This is a semianalytic theory with an accuracy, without differential correction, of about $\pm 0'' \cdot 005$ in longitude and latitude and ± 5 m in radial distance. These figures come from comparisons with the Jet Propulsion Laboratory (JPL) numerical integration, LE200. The precision is improved by about two orders of magnitude by differential correction of the coefficients using a numerical integration into which improved values of parameters have been introduced. The almanac offices, of necessity, are using lunar theories that contain all of the physics of the problem including, for example, the planetary perturbations and the nonsphericity of the earth and moon. Yet, even here, the precision of numerical integration is not equal to that of the lunar laser ranging. So the extraction of the time measurement from the differential correction of the lunar orbit becomes almost impossible.

Schmidt[6] has a semianalytical solution to the *main* problem[d] of the lunar theory. His solution offers a precision of ± 2 cm in the radial distance and ± 20 cm in the other directions. Instead of applying differential corrections, his coefficients are easily reevaluated with new values of the parameters in about 90 minutes.

None of these theories reach the precision levels required by timekeeping. Throughout history, lunar theories seem to have been cursed by finding themselves, just as they are completed, just below the level of precision needed for current observations. Delaunay[7] finished his lunar theory to eighth order in 1860 with a precision equivalent to about ± 40 km, but the observations were ahead of him. Continuing to work until 1869, three years before his death, he still had not succeeded in obtaining the precision necessary. E. W. Brown[8] proposed using Hill's method to accelerate the calculations of the lunar ephemeris, and published his solution to the main problem in 1908.[9] It appeared that he had caught up with the observations, but he had yet to add the planetary perturbations. In 1925, from Brown's lunar tables, which were published in 1919,[10] two students, Eckert and Brouwer, found that there was a discrepancy between the theory and the observations. The error in the theory was ± 4 km.

The situation remained this way until after World War II, when Eckert made an arrangement with Watson, President of the IBM Corporation, to place the newest IBM equipment at Columbia University, where Eckert was a member of the faculty. By improving the lunar theory of Brown, he would simultaneously be testing the latest equipment. From this work, Eckert, Jones, and Clark published the Improved Lunar

---

[d]The main problem is defined as a three-body problem with the orbit of the sun taken to be an ellipse about the earth–moon barycenter.

Ephemeris (I.L.E.)[11] in 1954. Since this theory did not yet meet the precision of Brown's tables, Eckert kept on with the work, publishing, 12 years later,[12] differential corrections that now gave the lunar ephemeris with an error of about ±1 km. Yet even after this and his continuing improvements until his death in 1972, Eckert never met the challenge of observational accuracy. His final work was completed and published posthumously by Gutzwiller and Schmidt[13] in the Astronomial Papers of the American Ephemeris.

So far, the work described had been semianalytic. In the early 1970s, Deprit, Henrard, and Rom[14] went back to Delaunay's original theory for the main problem, solving it by the method of Lie series, entirely in literal form on a computer. The solution is far too long, requiring 24 million bytes for its expression, but it has an accuracy of ±50 cm. For a while, the theory could match, even exceed, the precision of lunar laser ranging, but it did not contain all of the physics. And soon observational accuracy got ahead. It was at Deprit's urging in the late 1970s that Schmidt began his work on redoing the Hill–Brown lunar theory on the Amdahl computer at the University of Cincinnati. And we are back at the beginning of our story.

We should be able to learn lessons from history. One of these is the necessity of developing theories that can be extended to higher orders as long as the computer power is there and that can be easily adjusted to improved values of the parameters. Literal developments offer the most promise in this regard, even just for the ever present need for differential correction. Yet we also know that the main problem requires 24 million bytes in literal form, and *that* for a precision more than 100 times too large. The problem has to be simplified.

ARTIFICIAL SATELLITE THEORY

The main problem in artificial satellite theory is described by the Hamiltonian

$$\mathcal{H} = \frac{1}{2}\left(R^2 + \frac{\Theta^2}{r^2}\right) - \frac{\mu}{r}\left[1 - J_2\left(\frac{\alpha}{r}\right)^2 P_2(\sin\beta)\right], \tag{1}$$

with the following relations

$$\sin\beta = \sin\theta \sin I, \qquad \theta = f + g,$$

$$r\cos f = a(\cos E - e), \qquad r\sin f = a\eta\sin E, \qquad \eta = \sqrt{1 - e^2}$$

$$r = a(1 - e\cos E), \qquad \ell = E - e\sin E = \sqrt{\mu/a^3}(t - t_0),$$

and J_2 is a small parameter.

From these relations, it may be possible to trace the implicit dependence of the Hamiltonian on the mean anomaly, ℓ. In one of the early treatments of this problem, Brouwer[15] eliminated ℓ from the Hamiltonian, using a Poincaré–von Zeipel transformation. His transformation actually belongs to the category of Delaunay normalizations as described by Deprit.[16] An important point for our discussion is that for the first time, someone had succeeded in eliminating the mean anomaly without writing the Hamiltonian as an *explicit* function of ℓ. All previous work had written the Hamiltonian as a Fourier series in ℓ with coefficients that are infinite series in powers of the eccentricity. Brouwer succeeded in treating ℓ as an *implicit* variable in the algorithm for the Delaunay normalization. The theory worked well to first order; there was no reason to believe that it could not be carried to higher orders.

A few years later, Kozai[17] published an extension of Brouwer's theory to the second order. Even in closed form in the eccentricities, his theory required hundred of terms in the generating function. In fact, the calculations proved to be so long that Kozai resorted to the expedient omission of terms that he thought too small to be detected in the observations, Alas, observations have by now become so precise that his missing terms should, in fact, be reinstated. Fifteen years later, when Coffey and Deprit[2] published the third order of the normalization, the authors say that proceeding along the same lines as Brouwer had become prohibitive, even though the extension was being done with literal developments on a computer. It was clear that the problem had to be simplified before the third order of the normalization could be done. It was at this point that serious questions were raised as to whether or not Brouwer's implicit algorithm would succeed at higher orders.

The key to obtaining normalization of the main problem to the third and fourth order (still in closed form in the eccentricity) has been the elimination of the parallax.[3] This method turns out to be the first instance in Celestial Mechanics of what has since been called a *simplification*.[18] (In fact, the precise meaning now given to the word "simplification" was motivated by this transformation. We shall give this definition in the later section on Simplification vs. Normalization.) In the elimination of the parallax, Deprit takes a Hamiltonian of the form

$$\mathcal{H} = \mathcal{H}_0 + \frac{\Theta^2}{r^2} \sum_{n\geq 1} \sum_{m\geq 0} \frac{\epsilon^n}{n!} \left(\frac{a}{r}\right)^n (C_{n,m} \cos m\theta + S_{n,m} \sin m\theta), \qquad (2)$$

and transforms it into

$$\mathcal{H} = \mathcal{H}'_0 + \frac{\Theta'^2}{r'^2} \sum_{n\geq 1} \frac{\epsilon^n}{n!} \mathcal{H}'_n,$$

where $\mathcal{H}'_n$ does not depend on the mean anomaly. Since the final form of the Hamiltonian depends on $1/r'^2$, the mean anomaly is still present; Delaunay normalization is not achieved. But the dependence on ℓ in the reduced Hamiltonian is *restricted* to that term.

The best feature of this reduction is seen when one considers the number of terms needed for the third-order generator. By applying the elimination of the parallax first and then following it with a normalization of the reduced problem, one has achieved a reduction by 80% in the number of terms needed to normalize the original problem through third order. The comcomitant computer time saved is enormous. Certainly, given the constraints of available computer power, it seems safe to say that normalization to fourth order could not have been done without first applying the elimination of the parallax. When the normalization of $1/r'^2$ is performed, the generator will contain $f - \ell$, the equation of the center. The complexity of normalizing without first simplifying comes from the fact that the transformation is juggling Fourier series in the true anomaly coming not only from powers of r not equal to 2 and trigonometric functions of multiples of θ but also from powers of the equation of the center. The simplification isolated the equation of the center, which was at once recognized to be the most troublesome feature of the theory.

TWO OTHER EXAMPLES

In this section, we shall discuss two other dynamical systems for which other ways of simplifying have been suggested. The first is the Gylden problem, a Keplerian system with a time-dependent Gaussian parameter. Its Hamiltonian is

$$\mathcal{H} = \frac{1}{2}\|\mathbf{X}\|^2 - \frac{\mu(t)}{\|\mathbf{x}\|},$$

where x is the position vector and **X** its conjugate momentum. Using a time-dependent, linear transformation of the phase variables, together with a transformation of the independent variable, suggested by Deprit,[19] it is possible[20] to transform the Hamiltonian into one representing an ordinary Keplerian system, to which is added a perturbation proportional to the second derivative of the parameter. Visibly, the transformed Hamiltonian becomes

$$\mathcal{H}' = \frac{1}{2}\|\mathbf{X}'\|^2 - \frac{\mu_0}{\|\mathbf{x}'\|} - \left(\frac{2}{\mu(\tau)}\frac{d^2\mu}{d\tau^2}\right)\|\mathbf{x}'\|^2.$$

If μ varies adiabatically, then one may expand the perturbation in a power series in τ, thereby making the Hamiltonian conservative through second order.

By virtue of what is done here, one must admit that the original Hamiltonian has been simplified. In the adiabatic case, the explicit time dependence of the Hamiltonian has been suppressed until the third order. Also in reference 20, a sort of double normalization is given that removes τ and ℓ completely from the Hamiltonian to all orders. It imparts to $\mathcal{H}'$ two symmetries generated by T' and L' that are now integrals of the motion; but the normalization places this time dependence in the generator of the mapping. For this reason, although formally interesting, the transformation is valid for only a limited range of the independent variable.

The second example we wish to mention is the planetary theory. A recent first-order theory for the planar problem has been completed[21] that is done entirely in elliptic functions. The motivation for the research came from the Neptune–Pluto problem, where the reciprocal of the distance between the two planets, Δ^{-1}, can be very large. Consequently, the expansion of this term in the traditional way, as a power series in the ratio of the semimajor axes of the two orbits, leads to very long series. A precision of $0'' \cdot 001$ requires this parameter to the seventieth power. A simplified solution is obtained by Picard iteration, adopting circular orbits for the reference solution. The reference orbit approximation gives

$$\Delta = A\sqrt{1 - \kappa^2 \sin^2 \phi},$$

where A and κ are constants and ϕ is a linear function of the synodic angle. One can now see more clearly why it is that elliptic functions are a natural choice for the expression of the solution.

We call the technique of using elliptic functions instead of the more traditional trigonometric functions a simplification, because the perturbations, at first order, are given by very concise formulas. Often, they are expressed by a finite number of terms. And, in fact, far from being hampered by small divisors, resonant orbits with the frequency ratio $(2p - 1)/(2p + 1)$ are given by purely periodic expressions of finite length through first order.

Clearly, the choice of function in which to write the perturbation simplifies the solution and, at the resonance, gives us some hope that small divisors may indeed be an artifact of the method of solution.

SIMPLIFICATION vs. NORMALIZATION

After considering the several examples of simplifying techniques, it became clear that some of the ideas of simplification could be unified into one formalism. The examples taken from satellite theory and the Gylden problem are both Hamiltonian systems and both simplifications do not eliminate one variable completely, but reduce the number of terms containing that variable in the Hamiltonian. The simplification of the planetary theory does not fit this pattern. That may be a feature intrinsic to the problem, or is possibly caused by the fact that the problem is not formulated here in a Hamiltonian context. The problem needs to be studied more closely. Concentrating on Hamiltonian systems, and the satellite problem in particular, we can arrive at a general definition of a *simplification*. To see how this is done, consider the Hamiltonian system

$$\mathcal{H} = \mathcal{H}_0 + \mathcal{P} = \mathcal{H}_0 + \sum_{n\geq 1} \frac{\epsilon^n}{n!} \mathcal{H}_n,$$

where $\mathcal{H}_0$ represents the principal part, $\mathcal{P}$ the perturbation with ϵ a small parameter. We are looking for canonical transformations $\Psi : (\mathbf{x}, \mathbf{X}) \mapsto (\mathbf{x}', \mathbf{X}')$ given by the generating function

$$\mathcal{W} = \sum_{m \geq 0} \frac{\epsilon^m}{m!} \mathcal{W}_{m+1},$$

which maps $\mathcal{H} \to \mathcal{K}$, where

$$\mathcal{K} = \mathcal{K}_0 + \sum_{n\geq 1} \frac{\epsilon^n}{n!} \mathcal{K}_n,$$

with $\mathcal{K}_0 = \mathcal{H}_0$. As we are interested in Lie transformations,[22] at any order we face the partial differential relation

$$L_0(\mathcal{W}_n) + \mathcal{K}_n = \tilde{\mathcal{H}}_n,$$

where $\tilde{\mathcal{H}}_n$ is a function made up of $\mathcal{H}_n$ and Poisson brackets of terms derived in earlier stages of the transformation. The Hamiltonian $\mathcal{H}_0$ defines the Lie derivative, in terms of the Poisson bracket,

$$L_0 : f \mapsto L_0(f) = (f; \mathcal{H}_0).$$

The kernel of L_0, designated as $\ker(L_0)$, is the set of functions, f, such that $L_0(f) = 0$; the image of L_0, designated as $\text{Im}(L_0)$ is the set of functions f for which there is a function g such that $L_0(g) = f$.

Since we are dealing with perturbed Keplerian systems in this paper, we know that

$$\mathcal{H}_0 = \frac{1}{2} \|\mathbf{X}\|^2 - \frac{\mu}{\|\mathbf{x}\|},$$

or, in the variables of Delaunay,

$$\mathcal{H}_0 = -\frac{\mu^2}{2L^2}.$$

This latter expression for $\mathcal{H}_0$ leads to the following

$$L_0 = n \frac{\partial}{\partial \ell},$$

where $n = \mu^2/L^3$, making it easy to identify members of $\ker(L_0)$ as functions independent of the mean anomaly. A precise definition of a *simplification*, distinguished from a *normalization*, may be found in reference 18, where the authors use the Delaunay normalization and the elimination of the parallax as guides. A brief discussion follows.

We define a set $\mathcal{F}$ to be a Poisson algebra if it is a vector space of functions closed with respect to the operation of the Poisson bracket. A Delaunay normalization is a canonical transformation which maps perturbation functions in $\mathcal{F}$ to a particular subset of $\mathcal{F}$, namely, $\ker(L_0)$. A simplification is, in a sense, a less ambitious transformation that maps functions in $\mathcal{F}$ into a proper subspace of $\mathcal{F}$ that includes $\ker(L_0)$, but is not necessarily equal to it.

As an example, consider the elimination of the parallax in artificial satellite theory. From the Hamiltonian equation 2, we see that the perturbation belongs to the set

$$\mathcal{F} = \left\{ f : f = \frac{1}{r^n} (C_{n,m} \cos m\theta + S_{n,m} \sin m\theta) \right\},$$

where $C_{n,m}, S_{n,m} \in \ker(L_0)$. The elimination of the parallax maps functions in $\mathcal{F}$ into functions in the set

$$\mathcal{G} = \left\{ g : g = \frac{C_0}{r^2} \right\},$$

where $C_0 \in \ker(L_0)$. Clearly, $\mathcal{G} \subset \mathcal{F}$. Since the reduced Hamiltonian contains perturbation terms that are functions in $\mathcal{G}$, it still depends on the transformed mean anomaly, ℓ'. Thus the elimination of the parallax though not a normalization is instead a simplification according to the new definition.

THE RADIAL SIMPLIFICATION

The idea for the simplification we discuss in this section grew out of an attempt to find a transformation that would eliminate positive powers of r in the Hamiltonian in a manner analogous to what the elimination of the parallax does with $1/r^n$. Positive powers of r are expressed most conveniently by the eccentric anomaly, and so the problem was first addressed by seeking an algorithm to eliminate Fourier series in the eccentric anomaly. Series of this type do not form a Poisson algebra; the operation of the Poisson bracket on two representative functions from this set gives, as a result, Fourier series in the eccentric anomaly multiplied by 1 or by $1/r$. And so we began to examine the vector space of functions defined by the more general form

$$\frac{1}{r^p} \sum_{m \geq 0} (C_m \cos mE + S_m \sin mE).$$

It is hoped that the following discussion will make it clear that this set of functions is a subset of a Poisson algebra $\mathcal{R}$ such that

$$\mathcal{R} = \left\{ f : f = \frac{1}{r^i} \sum_{0 \leq j \leq k} \alpha_j r^j + \frac{R}{r^u} \sum_{0 \leq v \leq w} \beta_v r^v, \alpha_j, \beta_v \in \ker(L_0) \right\}, \tag{3}$$

where $\mathcal{R}$ is the momentum conjugate to r. It can be shown that $\mathcal{R}$ is a Poisson algebra. The set $\mathcal{R}$ includes most types of perturbations associated with Keplerian systems. For example, it contains the subspace of polynomials

$$\Sigma \alpha_{n_1, n_2, \ldots, n_6} x^{n_1} y^{n_2} z^{n_3} X^{n_4} Y^{n_5} Z^{n_6}, \qquad n_i \in \mathbf{N},$$

where $\mathbf{x}$ is a position vector, $\mathbf{X}$ its associated momentum, and the coefficients are in $\ker(L_0)$. To see this, consider that any component of the position vector $\mathbf{x}$, may be written, using the Whittaker map,

$$x_i = A_i r \cos \theta + B_i r \sin \theta, \qquad A_i, B_i \in \ker(L_0).$$

From the relations in the section on artificial satellite theory, together with the expression $r\dot{r} = rR = na^2 e \sin E$, derived from these, we see that

$$x_i = C_i + D_i r + E_i(rR), \qquad C_i, D_i, E_i \in \ker(L_0).$$

When x_i is raised to a power, the powers of rR that enter may be reduced to powers of r or powers of r multiplied by R, since R to an even power is an even function of E. After a similar treatment for $\mathbf{X}$, it may be established that the polynomial previously given is in $\mathcal{R}$.

Again, with the relations of the section on artificial satellite theory, it may be shown that

$$(\cos f)^k, (\sin f)^k \in \mathcal{R},$$

and with the Chebyshev polynomials, that

$$\cos kf, \sin kf \in \mathcal{R}.$$

The radial simplification got its name because it removes all positive powers of r and all but two negative powers from the Hamiltonian. More precisely, we quote the main result of Deprit,[23] who shows that it is possible to build a Lie transformation

$$\phi : (r', \theta', \nu', R', \Theta', N'; \epsilon) \mapsto (r, \theta, \nu, R, \Theta, N)$$

to convert the Hamiltonian

$$\mathcal{H} = \mathcal{H}_0 + \sum_{n\geq 1} \frac{\epsilon^n}{n!} \mathcal{H}_n, \quad \mathcal{H}_n \in \mathcal{R},$$

into the series

$$\mathcal{H}' = \mathcal{H}'_0 + \sum_{m\geq 1} \frac{\epsilon^m}{m!} \mathcal{H}'_m \tag{4}$$

where

$$\mathcal{H}_0 = \mathcal{H}'_0,$$

$$\mathcal{H}'_m = \gamma_m + \frac{\delta_m}{r'^2} + \frac{\zeta_m}{r'} R',$$

with γ_m, δ_m, and $\zeta_m \in \ker(L_0)$.

The construction consists in using, repeatedly if need be, the following identities:

$$r^m = \frac{(2m+1)a}{m+1}\left[1 - \frac{m}{2m+1}\frac{p}{r}\right]r^{m-1} - L_0\left(\frac{ar^{m+1}R}{(m+1)\mu}\right) \quad \text{for } m \geq 1,$$

$$\frac{1}{r} = \frac{1}{a} + L_0\left(\frac{rR}{\mu}\right),$$

$$\frac{1}{r^m} = \frac{2m-5}{(m-2)pr^{m-1}}\left[1 - \frac{(m-3)r}{(2m-5)a}\right] + L_0\left(\frac{R}{(m-2)\Theta^2 r^{m-3}}\right) \quad \text{for } m \geq 3,$$

$$r^m R = L_0\left(\frac{r^{m+1}}{m+1}\right) \quad \text{for } m \geq 0,$$

$$\frac{R}{r^m} = -L_0\left(\frac{1}{(m-1)r^{m-1}}\right) \quad \text{for } m \geq 2.$$

As the argument of L_0 is the generator of the transformation, we may also see that each term of the generator is an element of $\mathcal{R}$.

Judging from the simplicity of the Hamiltonian (4) compared to the perturbations of the general form of $\mathcal{R}$, it appears that the radial simplification may become a powerful tool in perturbation analysis. Consider the following list of perturbations $\subset \mathcal{R}$ that may be candidates for this remarkable technique:

Radiation pressure: $\epsilon \mathbf{r},$

Stark effect: $\epsilon \mathbf{k} \cdot \mathbf{r},$

Zeeman problem: $\epsilon \|\mathbf{k} \times \mathbf{r}\|^2,$

Relativistic Kepler: $\sum_{n\geq 0} \frac{1}{(n+1)!} \frac{1}{c^{2n}} \binom{1/2}{n+1} \left(\frac{\|\mathbf{X}\|}{m}\right)^{2n+2},$

Artificial satellite: $\quad \dfrac{\mu}{r} \sum_{n \geq 1} J_n \left(\dfrac{a}{r}\right)^n P_n(\sin \beta),$

Lunar theory: $\quad -\dfrac{\mu}{r'} \sum_{n \geq 1} a_n \left(\dfrac{r}{r'}\right)^n P_n(\cos S).$

This list is only a small part of the list of perturbed Keplerian systems that may be simplified by this technique. It means that the reduced Hamiltonian of all such problems will ultimately be expressed in the form

$$\mathcal{H} = \dfrac{\mu^2}{2L^2} + \alpha + \dfrac{\beta}{r^2} + \dfrac{\gamma}{r} R, \qquad \alpha, \beta, \gamma \in \ker(L_0).$$

Clearly, being able to classify a problem according to the values of these three parameters in $\ker(L_0)$ represents not only an incredible simplification but also opens up the possibility of establishing equivalences among many different types of physical problems.

The last remark to be made is that the normalization of the reduced Hamiltonian again obviously depends on which terms are present. The term α does not depend on ℓ, and so if this is the only term present, the radial simplification has acted like a normalization. The term β/r^2 produces perturbations proportional to the equation of the center, and the term $\gamma R/r$, those proportional to $\ell n/r$.

FINAL REMARKS

It is not completely clear whether or not the last item in the list of perturbations at the end of the preceding section, the lunar theory, should be included here, since the perturbation depends on r', the radius vector of the sun's orbit, as well as on the radius vector of the moon's orbit. Preliminary discussions have already started with S. Coffey, A. Deprit, and L. Healy to determine whether a radial simplification can be performed in the lunar theory. If the main problem of lunar theory is expressed in Jacobi coordinates, and if the orbit of the sun is assumed to be an ellipse, then the Hamiltonian in extended phase space consists of the Kepler term for the moon, $-\mu^2/2L^2$, a term $n'L'$ (the frequency and first Delaunay momentum of the sun), and a perturbation given by the entry in the list of the previous section. Since it seems possible to classify the Kepler term for the moon as zeroth order while treating the term $n'L'$ as first order, it may be that the Lie derivative depends only on the mean anomaly of the moon, ℓ. Therefore, the radial simplification with respect to the variable r should be possible. To see that more clearly, refer to the table of perturbations in the last section and note that $a_n \in \ker(L_0)$ and that $\cos S$ is of the form $A \cos f + B \cos f$, where $A, B \in \ker(L_0)$. Looking ahead, if all goes well, the reduced Hamiltonian would not depend on ℓ at all, and so we would have the case where the simplification is actually a normalization. Hopefully, then, this technique will effect a Delaunay normalization of the lunar theory without expansion in powers of the eccentricity of the moon. The answer as to whether or not a radial simplification with respect to r' is possible is being considered. If the answer is affirmative, we would be able to generate a solution to the main problem of the lunar theory also free from expansions in powers of the eccentricity of the sun's orbit.

ACKNOWLEDGMENTS

The generous help of André Deprit is gratefully acknowledged. The radial simplification is his transformation.

This paper is testimony to the generosity of the European Space Agency in granting a Giuseppe Colombo fellowship to one of the authors (S. F.). Thanks are also due to our colleague Dr. Bruce R. Miller at the National Bureau of Standards for his comments.

The authors wish to thank the National Bureau of Standards, where they were both visiting when this manuscript was prepared.

REFERENCES

1. DEPRIT, A. 1981. The main problem in the theory of artificial satellites to order four. J. Guid. Control **4**: 201–206.
2. COFFEY, S. & A. DEPRIT. 1982. Third-order solution to the main problem in satellite theory. J. Guid. Control **5**: 366–371.
3. DEPRIT, A. 1981. The elimination of the parallax in satellite theory. Celest. Mech. **24**: 111–153.
4. DEPRIT, A. 1988. A Lissajous transformation for elliptic oscillators. Submitted for publication in Celestial Mechanics.
5. CHAPRONT-TOUZE, M. & J. CHAPRONT. 1983. The lunar ephemeris ELP 2000. Astron. Astrophys. **124**: 50–62.
6. SCHMIDT, D. S. 1982. The main problem of lunar theory solved by the method of Brown. Celest. Mech. **26**: 75.
7. DELAUNAY, C. 1860. Theorie du Mouvement de la Lune, Vol. 1, Memoires de l'Academie des Sciences. Tome XXVIII, Paris.
8. BROWN, E. W. 1896. An Introductory Treatise on the Lunar Theory. Cambridge Univ. Press. London/New York.
9. BROWN, E. W. 1908. Theory of the Motion of the Moon, Memoirs of the Royal Astronomical Society, Part V: 1–103.
10. BROWN, E. W. & H. B. HEDRICK. 1919. Tables of the Motion of the Moon. Yale Univ. Press. New Haven.
11. ECKERT, W. J., R. JONES & H. K. CLARK. 1954. Construction of the lunar ephemeris. *In* Improved Lunar Ephemeris 1952–1954: 283–363. U.S. Govt. Printing Office. Washington, D.C.
12. ECKERT, W. J., M. J. WALKER & D. ECKERT. 1966. Transformations of the lunar coordinates and orbital parameters. Astron. J. **71**: 314–332.
13. GUTZWILLER, M. C. & D. S. SCHMIDT. 1986. The motion of the moon as computed by the method of Hill, Brown, and Eckert. Astronomical Papers XXIII (Part I). Nautical Almanac Office. U.S.N.O. Washington, D.C.
14. DEPRIT, A., J. HENRARD & A. ROM. 1971. Analytical lunar ephemeris: Delaunay's theory. Astron. J. **76**: 269–272.
15. BROUWER, D. 1959. Solution of the problem of artificial satellite theory without drag. Astron. J. **64**: 378–397.
16. DEPRIT, A., R. CUSHMAN & R. MOSAK. 1983. Normal form and representation theory. J. Math. Phys. **24**: 2102–2117.
17. KOZAI, Y. 1962. Second-order solution of artificial satellite theory without drag. Astron. J. **67**: 446–461.
18. DEPRIT, A. & S. FERRER. 1987. Simplifications in the theory of artificial satellites. Presented to the AAS/AIAA Astrodynamics Specialist Conf. Kalispell, Mont. AAS Paper 87-443.

19. DEPRIT, A. 1982. Secular accelerations in Gylden's problem. Celest. Mech. **31:** 1–22.
20. DEPRIT, A., B. R. MILLER & C. A. WILLIAMS. 1988. Gylden systems: Rotation of pericenters. Submitted for publication.
21. WILLIAMS, C. A., T. VAN FLANDERN & E. A. WRIGHT. 1987. First-order planetary perturbations with elliptic functions. Celest. Mech. **40:** 367–391.
22. DEPRIT, A. 1969. Canonical transformations depending on a small parameter. Celest. Mech. **1:** 12–30.
23. DEPRIT, A. 1987. Private communication.

Index of Contributors

Buchler, J. R., vii–viii

Chen, H. H., 91–99
Contopoulos, G., 1–14

Deprit, A., 101–126
de Zeeuw, T., 15–24
Dragt, A. J., 83–85

Ferrer, S., 127–139

Hietarinta, J., 33–42
Hunter, C., 25–32

Ipser, J. R. vii–viii, 77–80

Lin, J. E., 91–99
Littlejohn, R. G., 87–90
Lovelace, R. V. E., 81–82

Miller, B. R., 101–127

Shapiro, S. L., 53–75

Tabor, M., 43–51
Teukolsky, S. A., 53–75

Williams, C. A., vii–viii, 127–139